Lecture Notes in Mathematics 1629

Editors:
J.-M. Morel, Cachan
F. Takens, Groningen
B. Teissier, Paris

Springer
Berlin
Heidelberg
New York
Barcelona
Hong Kong
London
Milan
Paris
Singapore
Tokyo

John Douglas Moore

Lectures on
Seiberg-Witten Invariants

Second Edition

 Springer

Author

John Douglas Moore
Department of Mathematics
University of California
Santa Barbara, CA 93106, USA

E-mail: moore@math.ucsb.edu

Cataloging-in-Publication Data applied for

Die Deutsche Bibliothek - CIP-Einheitsaufnahme

Moore, John D.:
Lectures on Seiberg-Witten invariants / John Douglas Moore. - 2. ed..
- Berlin ; Heidelberg ; New York ; Barcelona ; Hong Kong ; London ;
Milan ; Paris ; Singapore ; Tokyo : Springer, 2001
  (Lecture notes in mathematics ; 1629)
  ISBN 3-540-41221-2

Mathematics Subject Classification (2000): 53C27, 58E15, 58J20

ISSN 0075-8434
ISBN 3-540-41221-2 Springer-Verlag Berlin Heidelberg New York

Springer-Verlag Berlin Heidelberg New York
a member of BertelsmannSpringer Science+Business Media GmbH

http://www.springer.de

© Springer-Verlag Berlin Heidelberg 2001
Printed in Germany

Typesetting: Camera-ready T$_E$X output by the authors
SPIN: 10701161   41/3142-543210 - Printed on acid-free paper

Preface

Riemannian, symplectic and complex geometry are often studied by means of solutions to systems of nonlinear differential equations, such as the equations of geodesics, minimal surfaces, pseudoholomorphic curves and Yang-Mills connections. For studying such equations, a new unified technology has been developed, involving analysis on infinite-dimensional manifolds.

A striking applications of the new technology is Donaldson's theory of "anti-self-dual" connections on $SU(2)$-bundles over four-manifolds, which applies the Yang-Mills equations from mathematical physics to shed light on the relationship between the classification of topological and smooth four-manifolds. This reverses the expected direction of application from topology to differential equations to mathematical physics. Even though the Yang-Mills equations are only mildly nonlinear, a prodigious amount of nonlinear analysis is necessary to fully understand the properties of the space of solutions.

At our present state of knowledge, understanding smooth structures on topological four-manifolds seems to require nonlinear as opposed to linear PDE's. It is therefore quite surprising that there is a set of PDE's which are even less nonlinear than the Yang-Mills equation, but can yield many of the most important results from Donaldson's theory. These are the Seiberg-Witten equations.

These lecture notes stem from a graduate course given at the University of California in Santa Barbara during the spring quarter of 1995. The objective was to make the Seiberg-Witten approach to Donaldson theory accessible to second-year graduate students who had already taken basic courses in differential geometry and algebraic topology.

In the meantime, more advanced expositions of Seiberg-Witten theory have appeared (notably [13] and [32]). It is hoped these notes will prepare the reader to understand the more advanced expositions and the excellent recent research literature.

We wish to thank the participants in the course, as well as Vincent Borrelli, Xianzhe Dai, Guofang Wei and Rick Ye for many helpful discussions on the material presented here.

J. D. MOORE
Santa Barbara
April, 1996

In the second edition, we have corrected several minor errors, and expanded several of the arguments to make them easier to follow. In particular, we included a new section on the Thom form, and provided a more detailed description of the second Stiefel-Whitney class and its relationship to the intersection form for four-manifolds. Even with these changes, the pace is demanding at times and increases throughout the text, particularly in the last chapter. The reader is encouraged to have pencil and paper handy to verify the calculations.

We have treated the Seiberg-Witten equations from the point of view of pure mathematics. The reader interested in the physical origins of the subject is encouraged to consult [9], especially the article, "Dynamics of quantum field theory," by Witten.

Our thanks go to David Bleecker for pointing out that our earlier proof of the Proposition on page 115 was incomplete, and to Lev Vertgeim and an anonymous referee for finding several misprints and minor errors in the text.

J. D. MOORE
Santa Barbara
February, 2001

Contents

Chapter 1

Preliminaries

1.1 Introduction

During the 1980's, Simon Donaldson utilized the Yang-Mills equations, which had originated in mathematical physics, to study the differential topology of four-manifolds. Using work of Michael Freedman, he was able to prove theorems of the following type:

Theorem A. *There exist many compact topological four-manifolds which have no smooth structure.*

Theorem B. *There exist many pairs of compact simply connected smooth four-manifolds which are homeomorphic but not diffeomorphic.*

The nonlinearity of the Yang-Mills equations presented difficulties, so many new techniques within the theory of nonlinear partial differential equations had to be developed. Donaldson's theory was elegant and beautiful, but the detailed proofs were difficult for beginning students to master.

In the fall of 1994, Edward Witten proposed a set of equations which give the main results of Donaldson theory in a far simpler way than had been thought possible. The purpose of these notes is to provide an elementary introduction to the equations which Witten proposed. These equations are now known as the *Seiberg-Witten equations*.

Our goal is to use the Seiberg-Witten equations to give the differential geometric parts of the proofs of Theorems A and B. The basic idea is simple: one constructs new invariants of smooth four-manifolds, invariants which depend upon the differentiable structure, not just the topology.

The reader is undoubtedly familiar with many topological invariants of four-manifolds: the fundamental group $\pi_1(M)$, the cohomology groups

$H^k(M)$, the cup product, and so forth. These topological invariants have been around for a long time and have been intensively studied. The Seiberg-Witten equations give rise to new invariants of four-dimensional smooth manifolds, called the *Seiberg-Witten invariants*. The key point is that homeomorphic smooth four-manifolds may have quite different Seiberg-Witten invariants. Just as homology groups have many applications, one might expect the Seiberg-Witten invariants to have many applications to the geometry and differential topology of four-dimensional manifolds.

Indeed, shortly after the Seiberg-Witten invariants were discovered, several striking applications were found.

One application concerns the geometry of embedded algebraic curves in the complex projective plane $\mathbb{C}P^2$. Any such curve has a degree, which is simply the number of times the curve intersects a projective line in general position.

Algebraic topologists have another way of calculating the degree. A nonsingular algebraic curve can be regarded as the image of a holomorphic embedding

$$i : \Sigma \to \mathbb{C}P^2,$$

Σ being a compact Riemann surface. The degree of the algebraic curve is the integer d such that

$$i_*(\text{fundamental class in } H_2(\Sigma; \mathbb{Z})) = d \cdot (\text{generator of } H_2(\mathbb{C}P^2; \mathbb{Z})). \quad (1.1)$$

In many algebraic geometry texts (for example, page 220 in [19]), one can find a formula for the genus of an embedded algebraic curve:

$$g = \frac{(d-1)(d-2)}{2}.$$

Thom conjectured that if Σ is a compact Riemann surface of genus g and

$$i : \Sigma \to \mathbb{C}P^2$$

is any smooth embedding, not necessarily holomorphic, then

$$g \geq \frac{(d-1)(d-2)}{2},$$

the degree being defined by (1.1). (One would not expect equality for general embeddings, since one can always increase the genus in a fixed homology class by adding small handles.)

The Thom conjecture was proven by Kronheimer and Mrowka, Morgan, Szabo and Taubes, and Fintushel and Stern, using the Seiberg-Witten equations. These notes should give the reader adequate background to read the proof (versions of which are presented in [23] and [33]). The proof also gives much new information about embeddings of surfaces in four-manifolds other than $\mathbb{C}P^2$.

Another application of the Seiberg-Witten invariants comes from differential geometry. One of the most studied problems in Riemannian geometry concerns the relationship between curvature and topology of Riemannian manifolds. Perhaps the simplest type of curvature is the scalar curvature

$$s : M \to \mathbb{R}$$

of a Riemannian manifold M. The value of the scalar curvature at p is a constant multiple of the average of all the sectional curvatures at p. It is interesting to ask: which compact simply connected Riemannian manifolds admit metrics with positive scalar curvature?

Lichnerowicz found the simplest obstruction to the existence of metrics of positive scalar curvature on compact simply connected manifolds. We will describe the part of Lichnerowicz's theorem that applies to four-manifolds later. Building upon the work of Lichnerowicz, Gromov and Lawson were able to obtain a relatively complete description of which compact simply connected manifolds of dimension ≥ 5 admit metrics of positive scalar curvature. (See [25], Corollary 4.5, page 301.)

As Witten noticed, a compact four-manifold with positive scalar curvature must have vanishing Seiberg-Witten invariants. Thus there is an obstruction to the existence of metrics of positive scalar curvature which depends on the differentiable structure of the four-manifold, not just its topological type. The Seiberg-Witten invariants show that many compact four-manifolds (including all compact algebraic surfaces of "general type") do not admit metrics of positive scalar curvature.

A third application of the Seiberg-Witten equations is to symplectic geometry. Indeed, Taubes [38] was able to identify the Seiberg-Witten invariants of a compact symplectic four-manifold with Gromov invariants— as a consequence he obtained an existence theorem for "pseudoholomorphic curves" in such manifolds.

The rapidity with which these new results have been obtained suggests that the Seiberg-Witten equations may have yet further applications to the geometry of four-manifolds. This is now an area of intensive research.

The differential geometry needed to study the Seiberg-Witten equations is the geometry of spin and spin^c structures. Until recently, these topics appeared unfamiliar and strange to many geometers, although spinors have long been regarded as important in physics. The tools needed to study spin and spin^c structures are the same standard tools needed by all geometers and topologists: vector bundles, connections, characteristic classes and so forth. We will begin by reviewing some of this necessary background.

1.2   What is a vector bundle?

Roughly speaking, a vector bundle is a family of vector spaces, parametrized by a smooth manifold M.

How does one construct such a family of vector spaces? Suppose first that the ground field is the reals and the vector spaces are to be of dimension m, all isomorphic to $\mathbb{R}^m$. In this case, one starts with an open covering $\{U_\alpha : \alpha \in A\}$ of M and for each $\alpha, \beta \in A$, smooth transition functions

$$g_{\alpha\beta} : U_\alpha \cap U_\beta \to GL(m, \mathbb{R}) = \{m \times m \text{ nonsingular real matrices}\},$$

which satisfy the "cocycle condition"

$$g_{\alpha\beta} \cdot g_{\beta\gamma} = g_{\alpha\gamma} \quad \text{on} \quad U_\alpha \cap U_\beta \cap U_\gamma.$$

Note that

$$g_{\alpha\alpha} \cdot g_{\alpha\beta} = g_{\alpha\beta} \quad \Rightarrow \quad g_{\alpha\alpha} = I \quad \text{on} \quad U_\alpha,$$

and hence the cocycle condition implies that

$$g_{\alpha\beta} \cdot g_{\beta\alpha} = g_{\alpha\alpha} = I \quad \text{on} \quad U_\alpha \cap U_\beta.$$

Let $\widetilde{E}$ denote the set of all triples $(\alpha, p, v) \in A \times M \times \mathbb{R}^m$ such that $p \in U_\alpha$. Define an equivalence relation $\sim$ on $\widetilde{E}$ by

$$(\alpha, p, v) \sim (\beta, q, w) \quad \Leftrightarrow \quad p = q \in U_\alpha \cap U_\beta, \quad v = g_{\alpha\beta}(p)w.$$

Denote the equivalence class of (α, p, v) by $[\alpha, p, v]$ and the set of equivalence classes by E. Define a projection map

$$\pi : E \to M \quad \text{by} \quad \pi([\alpha, p, v]) = p.$$

Let $\widetilde{U}_\alpha = \pi^{-1}(U_\alpha)$ and define a bijection

$$\psi_\alpha : \widetilde{U}_\alpha \to U_\alpha \times \mathbb{R}^m \quad \text{by} \quad \psi_\alpha([\alpha, p, v]) = (p, v).$$

There is a unique manifold structure on E which makes each ψ_α into a diffeomorphism. With respect to this manifold structure, the projection π is a smooth submersion.

A *real vector bundle of rank m over M* is a pair (E, π) constructed as above for some choice of open cover $\{U_\alpha : \alpha \in A\}$ of M and some collection $g_{\alpha\beta}$ of transition functions which satisfy the cocycle condition. The *fiber* of this vector bundle over $p \in M$ is $E_p = \pi^{-1}(p)$, the preimage of p under the projection. It has the structure of an m-dimensional real vector space.

When are two such vector bundles to be regarded as isomorphic? To answer this question, we need the notion of morphism within the category of vector bundles over M. A *vector bundle morphism* from (E_1, π_1) to (E_2, π_2) over M is a smooth map $f : E_1 \to E_2$ which takes the fiber $(E_1)_p$ of E_1 over p to the fiber $(E_2)_p$ of E_2 over p and restricts to a linear map on fibers, $f_p : (E_1)_p \to (E_2)_p$. An invertible vector bundle morphism is called a *vector bundle isomorphism* over M. Let $\mathrm{Vect}^{\mathbb{R}}_m(M)$ denote the space of isomorphism classes of real vector bundles of rank m over M.

The reader has no doubt encountered many examples of vector bundles in courses on differential geometry: the tangent bundle TM, the cotangent bundle T^*M, the k-th exterior power $\Lambda^k T^*M$ of the cotangent bundle, and other tensor bundles. Given two vector bundles E_1 and E_2 over M, one can form their direct sum $E_1 \oplus E_2$, their tensor product $E_1 \otimes E_2$, the bundle $\mathrm{Hom}(E_1, E_2)$, and so forth. One can also construct the dual bundle E_1^* whose fibers are the dual spaces to the fibers of E_1. The construction of such vector bundles is described in detail in §3 of [30].

Complex vector bundles are defined in a similar way. The only difference is that in the complex case the transition functions take their values in the group $GL(m, \mathbb{C})$ of $m \times m$ complex matrices instead of $GL(m, \mathbb{R})$, and $\widetilde{E}$ is replaced by the set of triples $(\alpha, p, v) \in A \times M \times \mathbb{C}^m$ such that $p \in U_\alpha$. The construction described above then gives a pair (E, π) in which the fiber $\pi^{-1}(p)$ has the structure of a *complex* vector space of dimension m. Let $\mathrm{Vect}^{\mathbb{C}}_m(M)$ denote the space of isomorphism classes of complex vector bundles of rank m over M.

A complex vector bundle of rank one is also called a *complex line bundle*. The space of complex line bundles forms an abelian group under the tensor product operation $\otimes$. We will sometimes write

$$L^m = L \otimes L \otimes \cdots \otimes L \qquad (m \text{ times}).$$

Note that if

$$g_{\alpha\beta} : U_\alpha \cap U_\beta \to GL(1, \mathbb{C})$$

are the transition functions for L, then the transition functions for L^m are simply $g_{\alpha\beta}^m$.

In addition to real and complex vector bundles, one can define quaternionic vector bundles, vector bundles over the quaternions. Quaternions were first described by William R. Hamilton in 1853. In modern notation, a quaternion is simply a 2×2 matrix of the form

$$Q = a\mathbf{1} + b\mathbf{i} + c\mathbf{j} + d\mathbf{k},$$

where a, b, c and d are real numbers, and

$$1 = \begin{pmatrix} 1 & 0 \\ 0 & 1 \end{pmatrix}, \quad \mathbf{i} = \begin{pmatrix} 0 & -1 \\ 1 & 0 \end{pmatrix}, \quad \mathbf{j} = \begin{pmatrix} 0 & i \\ i & 0 \end{pmatrix}, \quad \mathbf{k} = \begin{pmatrix} i & 0 \\ 0 & -i \end{pmatrix}.$$

It is readily checked that the sum of two quaternions or the product of two quarternions is again a quaternion. Quaternion multiplication is bilinear over the reals; thus it is determined by the multiplication table for its basis $\{1, \mathbf{i}, \mathbf{j}, \mathbf{k}\}$:

	1	i	j	k
1	1	i	j	k
i	i	−1	−k	j
j	j	k	−1	−i
k	k	−j	i	−1

Thus of two possible conventions, we choose the one which induces the *negative* of the cross product on the three-plane of "imaginary quaternions" spanned by $\mathbf{i}$, $\mathbf{j}$ and $\mathbf{k}$.

Alternatively, we can think of quaternions as 2×2 complex matrices of the form

$$\begin{pmatrix} w & -\bar{z} \\ z & \bar{w} \end{pmatrix},$$

where z and w are complex numbers. Note that since

$$\det Q = |z|^2 + |w|^2,$$

a nonzero quaternion Q possesses a multiplicative inverse.

We let $\mathbb{H}$ denote the space of quaternions. It is a skew field, satisfying all the axioms of a field except for commutativity of multiplication. Let $GL(m, \mathbb{H})$ denote the group of nonsingular $m \times m$ matrices with quaternion entries.

To define a quaternionic vector bundle of rank m, we simply require that the transition functions $g_{\alpha\beta}$ take their values in $GL(m, \mathbb{H})$. We let $\text{Vect}_m^{\mathbb{H}}(M)$ denote the space of isomorphism classes of quaternionic vector bundles of rank m over M.

Note that $GL(m, \mathbb{H})$ is a subgroup of $GL(2m, \mathbb{C})$, which in turn is a subgroup of $GL(4m, \mathbb{R})$. A quaternionic vector bundle of rank m can thought of as a real vector bundle of rank $4m$ whose transition functions $g_{\alpha\beta}$ take their values in $GL(m, \mathbb{H}) \subset GL(4m, \mathbb{R})$. More generally, if G is a Lie subgroup of $GL(m, \mathbb{R})$, a *G-vector bundle* is a rank m vector bundle whose transition functions take their values in G.

Let us suppose, for example, that G is the orthogonal group $O(m) \subset GL(m, \mathbb{R})$. In this case the transition functions of a G-vector bundle preserve the usual dot product on $\mathbb{R}^m$. Thus the bundle E inherits a *fiber metric*, a smooth function which assigns to each $p \in M$ an inner product

$$\langle\ ,\ \rangle_p : E_p \times E_p \to \mathbb{R}.$$

If G is the special orthogonal group $SO(n)$, a G-vector bundle possesses not only a fiber metric, but also an orientation.

Similarly, if G is the unitary group $U(m) \subset GL(m, \mathbb{C}) \subset GL(2m, \mathbb{R})$, a G-vector bundle is a complex vector bundle of rank m together with a *Hermitian metric*, a smooth function which assigns to each $p \in M$ a map

$$\langle\ ,\ \rangle_p : E_p \times E_p \to \mathbb{C}$$

which satisfies the axioms

1. $\langle v, w \rangle_p$ is complex linear in v and conjugate linear in w,

2. $\langle v, w \rangle_p = \overline{\langle w, v \rangle_p}$

3. $\langle v, v \rangle_p \geq 0$, with equality holding only if $v = 0$.

A *section* of a vector bundle (E, π) is a smooth map

$$\sigma : M \to E \qquad \text{such that} \qquad \pi \circ \sigma = \text{identity}.$$

If $\sigma \in \Gamma(E)$, the restriction of σ to U_α can be written in the form

$$\sigma(p) = [\alpha, p, \sigma_\alpha(p)], \qquad \text{where} \quad \sigma_\alpha : U_\alpha \to \begin{cases} \mathbb{R}^m \\ \mathbb{C}^m \\ \mathbb{H}^m \end{cases}$$

is a smooth map. The vector-valued functions σ_α are called the *local representatives* of σ and they are related to each other by the formula

$$\sigma_\alpha = g_{\alpha\beta}\sigma_\beta \qquad \text{on} \quad U_\alpha \cap U_\beta. \tag{1.2}$$

In the real or complex case, the set $\Gamma(E)$ of sections of E is a real or complex vector space respectively, and also a module over the space of smooth functions on M. In the quaternionic case, we need to be careful since quaternionic multiplication is not commutative. In this case, (1.2) shows that sections of E can be multiplied on the right by quaternions.

Example. We consider complex line bundles over the Riemann sphere S^2, regarded as the one-point compactification of the complex numbers, $S^2 =$

$\mathbb{C} \cup \{\infty\}$. Give $\mathbb{C}$ the standard complex coordinate z and let $U_0 = S^2 - \{\infty\}$, $U_\infty = S^2 - \{0\}$. For each integer n, define

$$g_{\infty 0} : U_\infty \cap U_0 \to GL(1, \mathbb{C}) \qquad \text{by} \qquad g_{\infty 0}(z) = \frac{1}{z^n}.$$

This choice of transition function defines a complex line bundle over S^2 which we denote by H^n. A section of H^n is represented by maps

$$\sigma_0 : U_0 \to \mathbb{C}, \qquad \sigma_\infty : U_\infty \to \mathbb{C}$$

such that

$$\sigma_\infty = \frac{1}{z^n} \sigma_0, \qquad \text{on} \quad U_\infty \cap U_0.$$

It can be proven that any complex line bundle over S^2 is isomorphic to H^n for some $n \in \mathbb{Z}$.

In particular, the cotangent bundle to S^2 must be isomorphic to H^n for some choice of n. A section σ of the cotangent bundle restricts to $\sigma_0 dz$ on U_0 for some choice of complex valued function σ_0. Over U_∞, we can use the coordinate $w = 1/z$, and write $\sigma = -\sigma_\infty dw$. Since $dw = -(1/z)^2 dz$,

$$\sigma_0 dz = -\sigma_\infty dw \qquad \Rightarrow \qquad \sigma_\infty = z^2 \sigma_0,$$

and hence $n = -2$. In other words, $T^* S^2 = H^{-2}$. Similarly, $T S^2 = H^2$.

In a similar way, we can construct all quaternionic line bundles over S^4. In this case, we regard S^4 as the one-point compactification of the space of quaternions, $S^4 = \mathbb{H} \cup \{\infty\}$. Let $U_0 = \mathbb{H}$, $U_\infty = S^4 - \{0\}$, and define

$$g_{\infty 0} : U_\infty \cap U_0 \to GL(1, \mathbb{H}) \qquad \text{by} \qquad g_{\infty 0}(Q) = \frac{1}{Q^n}.$$

As n ranges over the integers, we obtain all quaternionic line bundles over S^4.

How can we prove the claims made in the preceding paragraphs? Proofs can be based upon theorems from differential topology which classify vector bundles over manifolds. Here are two of the key results:

Classification Theorem for Complex Line Bundles. *If M is a smooth manifold, there is a bijection*

$$\mathrm{Vect}_1^{\mathbb{C}}(M) \cong H^2(M; \mathbb{Z}).$$

This theorem will be proven in §1.6. The theorem implies that

$$\mathrm{Vect}_1^{\mathbb{C}}(S^2) \cong H^2(S^2; \mathbb{Z}) \cong \mathbb{Z},$$

and we will see that H^m corresponds to $m \in \mathbb{Z}$ under the isomorphism. A argument similar to that for complex line bundles could be used to prove:

Classification Theorem for Quaternionic Line Bundles. *If M is a smooth manifold of dimension ≤ 4, there is a bijection*

$$\operatorname{Vect}_1^{\mathbb{H}}(M) \cong H^4(M; \mathbb{Z}).$$

1.3 What is a connection?

In contrast to differential topology, differential geometry is concerned with "geometric structures" on manifolds and vector bundles. One such structure is a connection. Evidence of the importance of connections is provided by the numerous definitions of connection which have been proposed.

A definition frequently used by differential geometers goes like this. Let

$$\chi(M) = \{\text{vector fields on } M\}, \qquad \Gamma(E) = \{\text{smooth sections of } E\}.$$

Definition 1. A *connection* on a vector bundle E is a map

$$\nabla^A : \chi(M) \times \Gamma(E) \to \Gamma(E)$$

which satisfies the following axioms (where $\nabla_X^A \sigma = \nabla^A(X, \sigma)$):

$$\nabla_X^A(f\sigma + \tau) = (Xf)\sigma + f\nabla_X^A\sigma + \nabla_X^A\tau, \tag{1.3}$$

$$\nabla_{fX+Y}^A\sigma = f\nabla_X^A\sigma + \nabla_Y^A\sigma. \tag{1.4}$$

Here f is a real-valued function if E is a real vector bundle, a complex-valued function if E is a complex vector bundle.

It is customary to regard $\nabla_X^A\sigma$ as the *covariant derivative* of σ in the direction of X.

Given a connection ∇^A in the sense of Definition 1, we can define a map

$$d_A : \Gamma(E) \to \Gamma(T^*M \otimes E) = \Gamma(\operatorname{Hom}(TM, E))$$

by

$$d_A(\sigma)(X) = \nabla_X^A\sigma.$$

Then d_A satisfies a second definition:

Definition 2. A *connection* on a vector bundle E is a map

$$d_A : \Gamma(E) \to \Gamma(T^*M \otimes E)$$

which satisfies the following axiom:

$$d_A(f\sigma + \tau) = (df) \otimes \sigma + f d_A\sigma + d_A\tau. \tag{1.5}$$

Definition 2 is more frequently used in gauge theory, but in our presentation both definitions will be important. Note that Definition 2 is a little more economical in that one need only remember one axiom instead of two. Moreover, Definition 2 makes clear the analogy between a connection and the exterior derivative.

The simplest example of a connection occurs on the bundle $E = M \times \mathbb{R}^m$, the trivial real vector bundle of rank m over M. A section of this bundle can be identified with a vector-valued map

$$\sigma = \begin{pmatrix} \sigma^1 \\ \sigma^2 \\ \cdot \\ \sigma^m \end{pmatrix} : M \to \mathbb{R}^m.$$

We can use the exterior derivative to define the "trivial" flat connection d_A on E:

$$d_A \begin{pmatrix} \sigma^1 \\ \cdot \\ \sigma^m \end{pmatrix} = \begin{pmatrix} d\sigma^1 \\ \cdot \\ d\sigma^m \end{pmatrix}.$$

More generally, given an $m \times m$ matrix

$$\omega = \begin{pmatrix} \omega_1^1 & \cdot & \omega_m^1 \\ \cdot & \cdots & \cdot \\ \omega_1^m & \cdot & \omega_m^m \end{pmatrix}$$

of real-valued one-forms, we can define a connection d_A by

$$d_A \begin{pmatrix} \sigma^1 \\ \cdot \\ \sigma^m \end{pmatrix} = \begin{pmatrix} d\sigma^1 \\ \cdot \\ d\sigma^m \end{pmatrix} + \begin{pmatrix} \omega_1^1 & \cdot & \omega_m^1 \\ \cdot & \cdots & \cdot \\ \omega_1^m & \cdot & \omega_m^m \end{pmatrix} \begin{pmatrix} \sigma^1 \\ \cdot \\ \sigma^m \end{pmatrix}. \tag{1.6}$$

We can write this last equation in a more abbreviated fashion:

$$d_A\sigma = d\sigma + \omega\sigma,$$

matrix multiplication being understood in the last term. Indeed, the axiom (1.5) can be verified directly, using the familiar properties of the exterior derivative:

$$d_A(f\sigma + \tau) = d(f\sigma + \tau) + \omega(f\sigma + \tau)$$

$$= df \otimes \sigma + f d\sigma + f\omega\sigma + d\tau + \omega\tau = (df) \otimes \sigma + f d_A \sigma + d_A \tau.$$

We can construct a connection in the trivial complex vector bundle $E = M \times \mathbb{C}^m$ or the trivial quaternionic bundle $M \times \mathbb{H}^m$ in exactly the same way, by choosing ω to be an $m \times m$ matrix of complex- or quaternionic-valued one-forms.

Any connection d_A on a trivial bundle is of the form (1.6). To see this, we apply d_A to the constant sections

$$e_1 = \begin{pmatrix} 1 \\ 0 \\ \cdot \\ 0 \end{pmatrix}, \quad e_2 = \begin{pmatrix} 0 \\ 1 \\ \cdot \\ 0 \end{pmatrix}, \quad \ldots \quad e_m = \begin{pmatrix} 0 \\ 0 \\ \cdot \\ 1 \end{pmatrix}.$$

obtaining

$$d_A e_j = \begin{pmatrix} \omega_j^1 \\ \omega_j^2 \\ \cdot \\ \omega_j^m \end{pmatrix} = \sum_{i=1}^m e_i \omega_j^i.$$

It then follows directly from the axiom for connections that

$$d_A \left(\sum_{i=1}^m e_i \sigma^i \right) = \sum_{i=1}^m e_i d\sigma^i + \sum_{i,j=1}^m e_i \omega_j^i \sigma^j,$$

which is just another way of writing (1.6).

Any vector bundle is "locally trivial." Suppose, for example, that E is a real vector of rank m over M defined by the open covering $\{U_\alpha : \alpha \in A\}$ and the transition functions $g_{\alpha\beta}$. In the notation of the preceding section

$$\psi_\alpha : \pi^{-1}(U_\alpha) \to U_\alpha \times \mathbb{R}^m$$

is a vector bundle isomorphism from the restriction of E to U_α onto the trivial bundle over U_α. A section $\sigma \in \Gamma(E)$ possesses a local representative σ_α, which is an m-tuple of ordinary functions on U_α. If d_A is a connection on E, then $d_A \sigma$ possesses a local representative $(d_A \sigma)_\alpha$, which is an m-tuple of ordinary one-forms on U_α. Just as in the preceding paragraph, we can write

$$(d_A \sigma)_\alpha = d\sigma_\alpha + \omega_\alpha \sigma_\alpha, \tag{1.7}$$

where ω_α is an $m \times m$ matrix of one-forms on U_α. This matrix ω_α is called the *local representative* of the connection d_A.

To see how the local representatives corresponding to two elements of our distinguished covering are related, we note that since the connection is well-defined on overlaps, we must have

$$d\sigma_\alpha + \omega_\alpha \sigma_\alpha = g_{\alpha\beta}(d\sigma_\beta + \omega_\beta \sigma_\beta) \quad \text{on} \quad U_\alpha \cap U_\beta.$$

Since $\sigma_\beta = g_{\alpha\beta}^{-1}\sigma_\alpha$,

$$d\sigma_\alpha + \omega_\alpha\sigma_\alpha = g_{\alpha\beta}[d(g_{\alpha\beta}^{-1}\sigma_\alpha) + \omega_\beta g_{\alpha\beta}^{-1}\sigma_\alpha]$$

$$= d\sigma_\alpha + g_{\alpha\beta}[dg_{\alpha\beta}^{-1} + g_{\alpha\beta}\omega_\beta g_{\alpha\beta}^{-1}]\sigma_\alpha.$$

Thus we conclude that

$$\omega_\alpha = g_{\alpha\beta}dg_{\alpha\beta}^{-1} + g_{\alpha\beta}\omega_\beta g_{\alpha\beta}^{-1} \qquad \text{on} \quad U_\alpha \cap U_\beta. \tag{1.8}$$

This yields

Definition 3. A *connection* on a real vector bundle E defined by a covering $\{U_\alpha : \alpha \in A\}$ and transition functions $\{g_{\alpha\beta} : \alpha, \beta \in A\}$ is a collection of differential operators

$$\{d + \omega_\alpha : \alpha \in A\},$$

where d is the exterior derivative on $\mathbb{R}^m$-valued functions, and ω_α is an $m \times m$ matrix of one-forms on U_α, which transform according to (1.8).

There is, of course, a similar definition for connections in complex or quaternionic vector bundles.

If E is an $O(m)$-bundle, an *orthogonal connection* in E is a connection whose local representatives $d+\omega_\alpha$ have the property that ω_α takes its values in $\mathcal{O}(m)$, where $\mathcal{O}(m)$ is the space of $m \times m$ skew-symmetric matrices, the Lie algebra of the orthogonal group $O(m)$. Note that this condition is preserved under the transformation (1.8) because

$$g_{\alpha\beta} \in O(m), \quad \omega_\beta \in \mathcal{O}(m) \qquad \Rightarrow \qquad g_{\alpha\beta}dg_{\alpha\beta}^{-1}, \quad g_{\alpha\beta}\omega_\beta g_{\alpha\beta}^{-1} \in \mathcal{O}(m).$$

Similarly, if E is a $U(m)$-bundle, a *unitary connection* in E is a connection such that ω_α takes its values in $\mathcal{U}(m)$, where $\mathcal{U}(m)$ is the space of $m \times m$ skew-Hermitian matrices, the Lie algebra of the unitary group $U(m)$. The general rule is that if E is a G-bundle, a G-connection is a connections whose local representatives ω_α take values in the Lie algebra of G.

Here are some examples of connections:

Example 1. Suppose that $E = TM$, the tangent bundle of a smooth manifold M which is embedded in $\mathbb{R}^N$. The trivial bundle $M \times \mathbb{R}^N$ is then a direct sum,

$$M \times \mathbb{R}^N = TM \oplus NM,$$

where NM is the normal bundle of M in $\mathbb{R}^N$, whose fiber at $p \in M$ is the set of all vectors in $\mathbb{R}^N$ perpendicular to M at p. On the trivial bundle $M \times \mathbb{R}^N$, we can take the trivial flat connection defined by the vector-valued exterior derivative d. If $\sigma \in \Gamma(E) \subset \Gamma(M \times \mathbb{R}^N)$, then $d\sigma \in \Gamma(T^*M \otimes (M \times \mathbb{R}^N))$. We define a connection

$$d_A : \Gamma(TM) \to \Gamma(T^*M \otimes TM)$$

by setting

$$d_A(\sigma) = (d\sigma)^\top,$$

where $(\cdot)^\top$ denotes projection into the tangent space. It is an enlightening exercise to check that this is just the Levi-Civita connection studied in Riemannian geometry.

Example 2. Over the Grassmann manifold

$$G_m(\mathbb{C}^N) = \{\ m\text{-dimensional subspaces of } \mathbb{C}^N\ \}$$

the trivial bundle divides into a direct sum,

$$G_m(\mathbb{C}^N) \times \mathbb{C}^N = E \oplus E^\perp,$$

where

$$E = \{(V, v) \in G_m(\mathbb{C}^N) \times \mathbb{C}^N : v \in V\},$$
$$E^\perp = \{(V, v) \in G_m(\mathbb{C}^N) \times \mathbb{C}^N : v \perp V\}.$$

As in the preceding example, we can use the trivial flat connection d on the trivial bundle $G_m(\mathbb{C}^N) \times \mathbb{C}^N$ to define a connection d_A on E by setting

$$d_A(\sigma) = (d\sigma)^\top,$$

where $(\cdot)^\top$ denotes the orthogonal projection into E. This connection is called the *universal connection* in the *universal bundle* E.

Example 3: the pullback connection. If (E, π) is a vector bundle over M defined by the open covering $\{U_\alpha : \alpha \in A\}$ and the transition functions $\{g_{\alpha\beta} : \alpha, \beta \in A\}$, and $F : N \to M$ is a smooth map, the *pullback bundle* (F^*E, π^*) is the vector bundle over N defined by the open covering $\{F^{-1}(U_\alpha) : \alpha \in A\}$ and the transition functions $\{g_{\alpha\beta} \circ F : \alpha, \beta \in A\}$. An alternate description is often useful:

$$F^*E = \{(p, v) \in N \times E : F(p) = \pi(v)\}, \qquad \pi^*((p, v)) = p.$$

If $\sigma \in \Gamma(E)$ has local representatives σ_α, we can define $F^*\sigma \in \Gamma(F^*E)$ to be the section with local representatives $\sigma_\alpha \circ F$. Equivalently, in terms of the alternate description,

$$F^*\sigma : N \to F^*E \qquad \text{by} \qquad F^*\sigma(p) = (p, \sigma \circ F(p)).$$

More generally, if $\omega \otimes \sigma$ is a differential form with values in E, we can define its pullback by

$$F^*(\omega \otimes \sigma) = F^*\omega \otimes F^*\sigma.$$

Proposition 1. *If d_A is a connection on the vector bundle (E, π) over M and $F : N \to M$ is a smooth map, there is a unique connection d_{F^*A} on the pullback bundle (F^*E, π^*) which makes the following diagram commute:*

$$\begin{array}{ccc} \Gamma(E) & \longrightarrow & \Gamma(T^*M \otimes E) \\ \downarrow & & \downarrow \\ \Gamma(F^*E) & \longrightarrow & \Gamma(T^*N \otimes F^*E) \end{array}$$

In this diagram, the horizontal arrows are given by the connections and the vertical arrows are the pullback maps F^*.

Proof: The commutative diagram implies that if ω_α is the local representative of d_A corresponding to U_α, then the local representative of d_{F^*A} corresponding to $F^{-1}(U_\alpha)$ is $F^*\omega_\alpha$. This establishes uniqueness. For existence, simply define the local representatives of d_{F^*A} to be $F^*\omega_\alpha$ and check that they satisfy the correct transformation formulae.

An important application of the pullback construction is to the existence of parallel transport along curves. If (E, π) is a smooth real vector bundle ov rank m over M and $\gamma : [a, b] \to M$ is a smooth curve, we can form the pullback bundle (γ^*E, π^*) over the interval $[a, b]$. It is an easy exercise to show that any vector bundle over an interval $[a, b]$ is trivial, so we can consider γ^*E to be the trivial bundle $[a, b] \times \mathbb{R}^m$. The pullback connection can be written as

$$d_{\gamma^*A} = d + \omega$$

where ω is a matrix of one-forms on $[a, b]$. If t is the standard coordinate on $[a, b]$,

$$\omega = \begin{pmatrix} f_1^1 dt & \cdot & f_m^1 dt \\ \cdot & & \cdot \\ f_1^m dt & \cdot & f_m^m dt \end{pmatrix},$$

where $f_j^i : [a, b] \to \mathbb{R}$. We now consider the equation

$$d_{\gamma^*A}\sigma = d\sigma + \omega\sigma = 0,$$

which can be written in terms of components as

$$\frac{d\sigma^i}{dt} + \sum_{j=1}^m f_j^i \sigma^j = 0, \tag{1.9}$$

a linear system of ordinary differential equations. It follows from the theory of differential equations that given an element $\sigma_0 \in (\gamma^* E)_a$, there is a unique solution to (1.9) which satisfies the initial condition $\sigma(a) = \sigma_0$. Thus we can define an isomorphism

$$\tau : (\gamma^* E)_a \to (\gamma^* E)_b$$

by setting $\tau(\sigma_0) = \sigma(b)$, where σ is the unique solution to (1.9) which satisfies $\sigma(a) = \sigma_0$. But

$$(\gamma^* E)_a \cong E_{\gamma(a)} \qquad \text{and} \qquad (\gamma^* E)_b \cong E_{\gamma(b)}.$$

Thus τ defines an isomorphism

$$\tau : E_{\gamma(a)} \to E_{\gamma(b)}$$

which we call *parallel transport along* γ.

It is not difficult to show that in the case of an $O(m)$-bundle with orthogonal connection, or a $U(m)$-bundle with unitary connection, parallel transport is an isometry.

Parallel transport has the following simple application:

Proposition 2. *If $F_0, F_1 : N \to M$ are smoothly homotopic maps and E is a vector bundle over M, then $F_0^* E$ and $F_1^* E$ are isomorphic vector bundles over N.*

Sketch of proof: Let $J_0, J_1 : N \to N \times [0, 1]$ be the smooth maps defined by

$$J_0(p) = (p, 0), \qquad J_1(p) = (p, 1).$$

If F_0 is smoothly homotopic to F_1, there exists a smooth map $H : N \times [0, 1] \to M$ such that

$$H \circ J_0 = F_0, \qquad H \circ J_1 = F_1.$$

Thus it suffices to show that if E is a vector bundle over $N \times [0, 1]$, then $J_0^* E$ is isomorphic to $J_1^* E$.

Give E a connection and let $\tau_p : E_{(p,0)} \to E_{(p,1)}$ denote parallel transport along the curve $t \mapsto (p, t)$. We can then define a vector bundle isomorphism $\tau : J_0^* E \to J_1^* E$ by

$$\tau(p, v) = (p, \tau_p(v)), \qquad \text{for} \qquad v \in E_{(p,0)} = J_0^* E_p.$$

Corollary. *Every vector bundle over a contractible manifold is trivial.*

Proof: If M is a contractible manifold, the identity map on M is homotopic to the constant map, and hence any vector bundle over M is isomorphic to the pullback of a bundle over a point via the constant map.

1.4   The curvature of a connection

Let's review some familiar facts regarding the exterior derivative. Let

$$\Omega^p(M) = \Gamma(\Lambda^p(T^*M)) = \{\text{differential } p\text{-forms on } M\},$$

so that $\Omega^0(M)$ is just the space of smooth real-valued functions on M. The exterior derivative $d : \Omega^0(M) \to \Omega^1(M)$ is just the usual differential of a function

$$df = \text{differential of } f = \sum_{i=1}^{n} \frac{\partial f}{\partial x^i} dx^i.$$

We extend this operator to $\Omega^p(M)$ for all p by requiring that $d(dx^i) = 0$ and that the Leibniz rule hold:

$$d(\omega \wedge \theta) = d\omega \wedge \theta + (-1)^p \omega \wedge d\theta, \qquad \text{for } \omega \in \Omega^p(M).$$

Equality of mixed partial derivatives implies that $d \circ d = 0$. Thus we obtain a cochain complex

$$\cdots \to \Omega^{p-1}(M) \to \Omega^p(M) \to \Omega^{p+1}(M) \to \cdots,$$

and we can define the p-th *de Rham cohomology group*

$$H^p(M; \mathbb{R}) = \frac{\text{kernel of } d : \Omega^p(M) \to \Omega^{p+1}(M)}{\text{image of } d : \Omega^{p-1}(M) \to \Omega^p(M)}.$$

It is well-known that these groups are isomorphic to the usual singular cohomology groups with real coefficients, and are therefore topological invariants. (Further discussion of de Rham theory can be found in [8].)

Now let E be a vector bundle over M and let

$$\Omega^p(E) = \Gamma(\Lambda^p(T^*M) \otimes E) = \{p\text{-forms on } M \text{ with values in } E\}.$$

A connection on E can be regarded as a linear map from zero-forms with values in E to one-forms with values in E,

$$d_A : \Omega^0(E) \to \Omega^1(E).$$

As in the case of the usual exterior derivative, we can extend d_A to all of the $\Omega^p(E)$'s by requiring Leibniz's rule to hold,

$$d_A(\omega\sigma) = d\omega \otimes \sigma + (-1)^p \omega \wedge d_A\sigma, \qquad \text{for } \omega \in \Omega^p(M), \quad \sigma \in \Gamma(E).$$

It is not usually true that $d_A \circ d_A = 0$. However,

$$d_A \circ d_A(f\sigma + \tau) = d_A[df \otimes \sigma + f d_A \sigma + d_A \tau]$$

$$= d(df)\sigma - df \wedge d_A\sigma + df \wedge d_A\sigma + f(d_A \circ d_A)\sigma + d_A \circ d_A\tau$$

$$= f(d_A \circ d_A\sigma) + d_A \circ d_A\tau,$$

so $d_A \circ d_A$ is linear over functions. This implies that $d_A \circ d_A$ is actually a tensor field, called the curvature of the connection.

In the case of the trivial vector bundle $E = M \times \mathbb{R}^m$, we can apply $d_A \circ d_A$ to the standard constant sections

$$e_1 = \begin{pmatrix} 1 \\ 0 \\ \cdot \\ 0 \end{pmatrix}, \quad e_2 = \begin{pmatrix} 0 \\ 1 \\ \cdot \\ 0 \end{pmatrix}, \quad \dots, \quad e_m = \begin{pmatrix} 0 \\ 0 \\ \cdot \\ 1 \end{pmatrix}.$$

Then

$$d_A \circ d_A(e_i) = \begin{pmatrix} \Omega_i^1 \\ \Omega_i^2 \\ \cdot \\ \Omega_i^m \end{pmatrix} = \sum_{j=1}^m e_j \Omega_i^j,$$

where each Ω_i^j is a two-form. Linearity over functions implies that

$$d_A \circ d_A \left(\sum_{i=1}^m e_i \sigma^i \right) = \sum_{i,j=1}^m e_i \Omega_j^i \sigma^j,$$

or in matrix notation

$$d_A \circ d_A(\sigma) = \Omega\sigma, \tag{1.10}$$

where Ω is a matrix of two-forms.

Suppose now that E is a real vector bundle of rank m over M defined by the open covering $\{U_\alpha : \alpha \in A\}$ and the transition functions $g_{\alpha\beta}$. Any element $\sigma \in \Gamma(E)$ possesses local representatives σ_α, and in accordance with (1.10),

$$(d_A \circ d_A\sigma)_\alpha = \Omega_\alpha\sigma_\alpha,$$

where Ω_α is a matrix of two-forms. Since d_A is represented by the operator $d+\omega_\alpha$ on $\Omega^0(E)$ and $d+\omega_\alpha$ satisfies the Leibniz rule, d_A must be represented by $d + \omega_\alpha$ on $\Omega^p(E)$ for all p. We therefore conclude that

$$(d_A \circ d_A\sigma)_\alpha = (d + \omega_\alpha)(d\sigma_\alpha + \omega_\alpha\sigma_\alpha)$$

$$= d(d\sigma_\alpha) + (d\omega_\alpha)\sigma_\alpha - \omega_\alpha \wedge d\sigma_\alpha + \omega_\alpha \wedge d\sigma_\alpha + (\omega_\alpha \wedge \omega_\alpha)\sigma_\alpha$$

$$= (d\omega_\alpha + \omega_\alpha \wedge \omega_\alpha)\sigma_\alpha,$$

and hence

$$\Omega_\alpha = d\omega_\alpha + \omega_\alpha \wedge \omega_\alpha. \tag{1.11}$$

Since $d_A \circ d_A$ is independent of trivialization, the matrices of two-forms must satisfy

$$\Omega_\alpha \sigma_\alpha = g_{\alpha\beta}\Omega_\beta \sigma_\beta = g_{\alpha\beta}\Omega_\beta g_{\alpha\beta}^{-1}\sigma_\alpha \qquad \text{on} \quad U_\alpha \cap U_\beta.$$

Thus the Ω_α's transform by the rule

$$\Omega_\alpha = g_{\alpha\beta}\Omega_\beta g_{\alpha\beta}^{-1}, \qquad\qquad (1.12)$$

the way in which local representatives of a two-form with values in the vector bundle $\operatorname{End}(E)$ should transform. In other words, the Ω_α's determine an $\operatorname{End}(E)$-valued two-form, called the *curvature* of the connection d_A, and sometimes denoted simply by Ω or Ω_A.

Differentiation of equation (1.11) yields the *Bianchi identity*:

$$d\Omega_\alpha = \Omega_\alpha \wedge \omega_\alpha - \omega_\alpha \wedge \Omega_\alpha = [\Omega_\alpha, \omega_\alpha]. \qquad\qquad (1.13)$$

Of course, all of the above discussion applies to connections in complex vector bundles or quaternionic vector bundles as well as real vector bundles. In the complex case, for example, the matrices Ω_α have complex-valued two-forms as entries.

In the case of an orthogonal connection in an $O(m)$-bundle, it follows directly from (1.11) that the matrices Ω_α are skew-symmetric. In the case of a unitary connection in a $U(m)$-bundle, these matrices are skew-Hermitian.

The case of complex line bundles is particularly important. In this case, the matrices Ω_α are all 1×1 and the transformation formula (1.12) implies that $\Omega_\alpha = \Omega_\beta$ on overlaps. Thus the Ω_α's fit together to make a globally defined two-form Ω_A on M, which is purely imaginary in the case of a unitary connection.

Since the Lie group $U(1)$ is isomorphic to $SO(2)$, a complex line bundle with Hermitian metric can also be regarded as a real vector bundle of rank two together with a fiber metric and orientation. Indeed, given an oriented rank two real vector bundle E with fiber metric, we can define multiplication by i to be rotation through 90 degrees in the direction determined by the orientation. If e_1 is a locally defined unit-length section of E, we can set $e_2 = ie_1$, thereby obtaining a locally defined moving frame (e_1, e_2). Note that if ω_{12} is the corresponding real valued connection form,

$$de_1 = -e_2\omega_{12} = (-i\omega_{12})e_1,$$

and $-i\omega_{12}$ is the purely imaginary connection form for the unitary bundle. We set

$$F_A = \Omega_{12} = d\omega_{12},$$

a globally defined curvature two-form on M. Then $\Omega_A = -iF_A$ is the corresponding purely imaginary two-form when E is regarded as a unitary bundle. The Bianchi identity in this case is simply

$$dF_A = 0. \tag{1.14}$$

All of this applies to the case where $E = T\Sigma$, the tangent bundle of an oriented two-dimensional Riemannian manifold Σ. In this case the Levi-Civita connection on $T\Sigma$ is a unitary connection and its curvature form is

$$F_A = KdA,$$

where K is the Gaussian curvature of Σ and dA is the area two-form of Σ.

1.5 Characteristic classes

In general, the curvature matrix Ω is only locally defined. However, it is possible to construct certain polynomials in Ω_α which are invariant under the transformation (1.12) and these give rise to geometric and topological invariants.

Let us focus first on the case of a $U(m)$-vector bundle E, a complex vector bundle of rank m with Hermitian metric. We give E a unitary connection which has local representatives $d + \omega_\alpha$ and curvature matrices Ω_α. As we saw in the preceding section, Ω_α is skew-Hermitian, so

$$\frac{i}{2\pi}\Omega_\alpha, \qquad \left(\frac{i}{2\pi}\Omega_\alpha\right)^k$$

are Hermitian. Hence for each choice of positive integer k, the differential form

$$\text{Trace}\left[\left(\frac{i}{2\pi}\Omega_\alpha\right)^k\right]$$

is real-valued. Since trace is invariant under similarity, it follows from (1.12) that

$$\text{Trace}\left[\left(\frac{i}{2\pi}\Omega_\alpha\right)^k\right] = \text{Trace}\left[\left(\frac{i}{2\pi}g_{\alpha\beta}\Omega_\beta g_{\alpha\beta}^{-1}\right)^k\right] = \text{Trace}\left[\left(\frac{i}{2\pi}\Omega_\beta\right)^k\right]$$

on $U_\alpha \cap U_\beta$. Hence these locally defined forms fit together into a globally defined real-valued $2k$-form $\tau_k(A)$. We say that $\tau_k(A)$ is a *characteristic form*.

Lemma. *The differential form $\tau_k(A)$ is closed, $d\tau_k(A) = 0$.*

Proof: It follows from the Bianchi identity that

$$d\tau_k(A) = \left(\frac{i}{2\pi}\right)^k d[\text{Trace}(\Omega_\alpha^k)] = \left(\frac{i}{2\pi}\right)^k \text{Trace}[d(\Omega_\alpha^k)]$$

$$= \left(\frac{i}{2\pi}\right)^k \text{Trace}[(d\Omega_\alpha)\Omega_\alpha^{k-1} + \cdots + \Omega_\alpha^{k-1}(d\Omega_\alpha)]$$

$$= \left(\frac{i}{2\pi}\right)^k \text{Trace}\{[\Omega_\alpha, \omega_\alpha]\Omega_\alpha^{k-1} + \cdots + \Omega_\alpha^{k-1}[\Omega_\alpha, \omega_\alpha]\} = 0,$$

the last equality coming from the fact that $\text{Trace}(A_1 \cdots A_k)$ is invariant under cyclic permutation of $A_1, \ldots, A_k$.

This Lemma implies that $\tau_k(A)$ represents a de Rham cohomology class

$$[\tau_k(A)] \in H^{2k}(M; \mathbb{R}).$$

Proposition 1. *If E is a $U(m)$-bundle over M with unitary connection d_A and $F : N \to M$ is a smooth map, then the characteristic forms of the pullback connection d_{F^*A} on the pullback bundle F^*E are pullbacks of the characteristic forms of d_A:*

$$\tau_k(F^*A) = F^*\tau_k(A).$$

Proof: The local representative of d_{F^*A} on $F^{-1}(U_\alpha)$ is $\omega_\alpha^* = F^*\omega_\alpha$, and the curvature of d_{F^*A} is

$$\Omega_\alpha^* = d\omega_\alpha^* + \omega_\alpha^* \wedge \omega_\alpha^* = F^*d\omega_\alpha + F^*\omega_\alpha \wedge F^*\omega_\alpha = F^*\Omega_\alpha.$$

Thus

$$\text{Trace}((\Omega_\alpha^*)^k) = \text{Trace}(F^*(\Omega_\alpha^k)) = F^*\text{Trace}(\Omega_\alpha^k),$$

or equivalently,

$$\tau_k(F^*A) = F^*\tau_k(A).$$

Proposition 2. *The de Rham cohomology class $[\tau_k(A)]$ is independent of the choice of unitary connection d_A as well as the choice of Hermitian metric on E.*

Proof: We use the cylinder construction from topology. From the bundle (E, π) over M we construct the cylinder bundle $(E \times [0, 1], \pi \times \text{id})$ over $M \times [0, 1]$. Note that if E possesses the trivializing covering $\{U_\alpha : \alpha \in A\}$, then $E \times [0, 1]$ possesses the trivializing covering $\{U_\alpha \times [0, 1] : \alpha \in A\}$.

We can pull two connections d_A and d_B on E back to $M \times [0,1]$ via the projection on the first factor $\pi_1 : M \times [0,1] \to M$. If d_A and d_B are given over U_α by

$$d + \omega_\alpha, \qquad d + \phi_\alpha,$$

then the pullback connections are given over $U_\alpha \times [0,1]$ by

$$d + \pi_1^* \omega_\alpha, \qquad d + \pi_1^* \phi_\alpha.$$

We define a new connection d_C on $E \times [0,1]$, given over $U_\alpha \times [0,1]$ by

$$d + (1-t)\pi_1^* \omega_\alpha + t\pi_1^* \phi_\alpha.$$

If $J_0, J_1 : M \to M \times [0,1]$ are the maps defined by $J_0(p) = (p,0)$, $J_1(p) = (p,1)$, then

$$d_A = d_{J_0^* C}, \qquad d_B = d_{J_1^* C}.$$

Hence by Proposition 1,

$$[\tau_k(A)] = J_0^*[\tau_k(C)] = J_1^*[\tau_k(C)] = [\tau_k(B)] \in H^{2k}(M; \mathbb{R}).$$

This shows that $[\tau_k(A)]$ is independent of the choice of unitary connection.

The proof that it is independent of the choice of Hermitian metric is similar and is left to the reader. Use the fact that any two Hermitian metrics $\langle , \rangle_0$ and $\langle , \rangle_1$ can be connected by a one-parameter family

$$(1-t)\langle \; , \; \rangle_0 + t\langle \; , \; \rangle_1.$$

According to Proposition 2, to each complex vector bundle E over M we can associate a collection of cohomology classes

$$\tau_k(E) = [\tau_k(A)].$$

These are called *characteristic classes*. Clearly, if two complex vector bundles are isomorphic over M, they must have the same characteristic classes. Characteristic classes are *natural* under mappings as the following proposition shows:

Proposition 3. *If* $F : N \to M$ *is a smooth map and* E *is a complex vector bundle over* M, *then*

$$\tau_k(F^* E) = F^* \tau_k(E).$$

Proof: This is now an immediate consequence of Proposition 1.

We can put the characteristic classes together into a power series called the *Chern character:*

$$\text{ch}(E) = [\text{Trace}\,(\exp\,((i/2\pi)\Omega_\alpha))]$$

$$= \text{rank}(E) + \tau_1(E) + \frac{1}{2!}\tau_2(E) + \cdots + \frac{1}{k!}\tau_k(E) + \cdots.$$

This is an element of the cohomology ring

$$H^*(M;\mathbb{R}) = H^0(M;\mathbb{R}) \oplus H^1(M;\mathbb{R}) \oplus \cdots \oplus H^k(M;\mathbb{R}) \oplus \cdots.$$

Note that the Chern character collapses to a polynomial since all terms of degree $> \dim(M)$ must vanish.

Proposition 4. *The Chern character satisfies the identities:*

$$\text{ch}(E_1 \oplus E_2) = \text{ch}(E_1) + \text{ch}(E_2), \qquad \text{ch}(E_1 \otimes E_2) = \text{ch}(E_1)\text{ch}(E_2).$$

Proof: Suppose that E_1 and E_2 have Hermitian metrics and unitary connections d_{A_1} and d_{A_2}, respectively. Then $E_1 \oplus E_2$ inherits a Hermitian metric and a unitary connection $d_{A_1 \oplus A_2}$, defined by

$$d_{A_1 \oplus A_2}(\sigma_1 \oplus \sigma_2) = (d_{A_1}\sigma_1) \oplus (d_{A_2}\sigma_2).$$

It follows that

$$(d_{A_1 \otimes A_2})^2(\sigma_1 \otimes \sigma_2) = (d_{A_1}^2\sigma_1) \oplus (d_{A_2}^2\sigma_2),$$

and hence

$$((i/2\pi)\Omega^{A_1 \oplus A_2})^k = \begin{pmatrix} ((i/2\pi)\Omega^{A_1})^k & 0 \\ 0 & ((i/2\pi)\Omega^{A_2})^k \end{pmatrix}.$$

Thus

$$\tau_k(E_1 \oplus E_2) = \tau_k(E_1) + \tau_k(E_2),$$

which implies the first of the two identities.

Similarly, $E_1 \otimes E_2$ inherits a Hermitian metric and a unitary connection $d_{A_1 \otimes A_2}$, defined by the Leibniz rule

$$d_{A_1 \otimes A_2}(\sigma_1 \otimes \sigma_2) = (d_{A_1}\sigma_1) \otimes \sigma_2 + \sigma_1 \otimes (d_{A_2}\sigma_2).$$

This rule shows that

$$(d_{A_1 \otimes A_2})^2(\sigma_1 \otimes \sigma_2) = (d_{A_1}^2\sigma_1) \otimes \sigma_2 + \sigma_1 \otimes (d_{A_2}^2\sigma_2),$$

from which it follows that

$$(d_{A_1 \otimes A_2})^{2k}(\sigma_1 \otimes \sigma_2) = \sum_{j=0}^{k} \binom{k}{j}(d_{A_1}^{2j}\sigma_1) \otimes (d_{A_2}^{2k-2j}\sigma_2),$$

or

$$\frac{1}{k!}(d_{A_1 \otimes A_2})^{2k}(\sigma_1 \otimes \sigma_2) = \sum_{j=0}^{k} \frac{1}{j!}(d_{A_1}^{2j}\sigma_1) \otimes \frac{1}{(k-j)!}(d_{A_2}^{2k-2j}\sigma_2).$$

We conclude that

$$\exp((d_{A_1 \otimes A_2})^2) = \exp((d_{A_1})^2)\exp((d_{A_2})^2),$$

which yields the second of the two identities.

In the topology of four-manifolds, certain polynomials in the $\tau_k(E)$'s play a key role. If E is a $U(m)$-vector bundle over M, the *first and second Chern classes* of E are defined by the formulae

$$c_1(E) = \tau_1(E), \qquad c_2(E) = \frac{1}{2}[\tau_1(E)^2 - \tau_2(E)]. \tag{1.15}$$

To motivate the second of these, we can check that if E has rank two, so that $(i/2\pi)\Omega_\alpha$ is a (2×2)-Hermitian matrix,

$$\frac{1}{2}\left[\operatorname{Trace}\left(\frac{i}{2\pi}\Omega_\alpha\right)^2 - \left(\operatorname{Trace}\frac{i}{2\pi}\Omega_\alpha\right)^2\right] = \operatorname{Determinant}\left(\frac{i}{2\pi}\Omega_\alpha\right),$$

so the second Chern class is also given by the formula

$$c_2(E) = \text{the de Rham cohomology class of Determinant}\left(\frac{i}{2\pi}\Omega_\alpha\right).$$

Higher Chern classes can also be defined by similar formulae, but they automatically vanish on manifolds of dimension ≤ 4.

Use of a Hermitian metric shows that the dual E^* to a $U(m)$-vector bundle E over M is obtained from E by conjugating transition functions,

$$g_{\alpha\beta} \mapsto \bar{g}_{\alpha\beta}.$$

Thus we can also think of E^* as the conjugate of E. A connection on E defines a connection on E^* by conjugating local representatives,

$$\omega_\alpha \mapsto \bar{\omega}_\alpha, \qquad \Omega_\alpha \mapsto \bar{\Omega}_\alpha.$$

Since conjugation changes the sign of the trace of a skew-Hermitian matrix, we see that $c_1(E^*) = -c_1(E)$; a similar argument shows that $c_2(E^*) = c_2(E)$.

We can also define characteristic classes of quaternionic and real vector bundles. A quaternionic line bundle E can be regarded as a complex vector bundle of rank two. Right multiplication by $\mathbf{j}$ is a conjugate-linear isomorphism from E to itself, so the first Chern class of E must vanish, but second Chern class is an important invariant of E.

On the other hand, a real vector bundle E of rank k can be complexified, giving a complex vector bundle $E \otimes \mathbb{C}$ of rank k. Once again, the complexification is isomorphic to its conjugate so its first Chern class must vanish, but we can define the *first Pontrjagin class* of E to be

$$p_1(E) = -c_2(E \otimes \mathbb{C}).$$

1.6   The Thom form

The first Chern class of a complex line bundle over a compact oriented surface possesses an important geometric interpretation as the algebraic number of zeros of a generic section. We now present that interpretation based upon the construction of a Thom form.

Suppose that L is a complex line bundle over M, regarded as an oriented real vector bundle of rank two, as described at the end of §1.4. If U is an open subset of M over which we have constructed a moving frame (e_1, e_2) with $e_2 = ie_1$, we can define smooth functions

$$t_1, t_2 : \pi^{-1}(U) \longrightarrow \mathbb{R} \quad \text{by} \quad t_i(a_1 e_1(p) + a_2 e_2(p)) = a_i.$$

Moreover, we can pull the connection and curvature forms ω_{12} and Ω_{12} back to $\pi^{-1}(U)$.

Definition. The *Thom form* on L is the smooth two-form defined locally by

$$\Phi = \frac{1}{2\pi} e^{-(t_1^2 + t_2^2)} \left[\Omega_{12} + 2(dt_1 + \omega_{12} t_2) \wedge (dt_2 - \omega_{12} t_1) \right]. \tag{1.16}$$

(It is not difficult to verify that this expression is independent of choice of moving frame (e_1, e_2), and hence Φ is indeed a globally defined differential two-form on L.)

We can also write

$$\Phi = \frac{1}{2\pi} e^{-(t_1^2 + t_2^2)} \left[\Omega_{12} + 2dt_1 \wedge dt_2 + \omega_{12} \wedge d(t_1^2 + t_2^2) \right],$$

from which it is easy to check that Φ is closed. Moreover, it is clearly rapidly decreasing in the fiber direction, and the familiar integral formula

$$\int_{R^2} e^{-(t_1^2+t_2^2)} dt_1 dt_2 = \int_0^\infty \int_0^{2\pi} e^{-r^2} r \, dr \, d\theta = \pi, \qquad (1.17)$$

shows that the integral of Φ over any fiber is one.

We reiterate the three key properties: (i.) The Thom form is closed. (ii.) It is rapidly decreasing in the fiber direction. (iii.) Its integral over any fiber is one. Any other two-form with with the same key properties would be equally effective in establishing a geometric interpretation for the first Chern class. Indeed, we could replace (1.16) with

$$\Phi = \eta(t_1^2 + t_2^2)\Omega_{12} - 2\eta'(t_1^2 + t_2^2)(dt_1 + \omega_{12}t_2) \wedge (dt_2 - \omega_{12}t_1),$$

where $\eta : [0, \infty) \to \mathbb{R}$ is any smooth function such that

$$\eta(0) = \frac{1}{2\pi}, \qquad \eta(u) \to 0 \text{ sufficiently rapidly as } u \to \infty,$$

thereby still obtaining a smooth two-form with the same key properties. The choice $\eta(u) = e^{-u}$ we made in (1.16) may simplify the calculations a little, but choosing η to vanish for $r \geq \epsilon$, where ϵ is a small positive number, yields a Thom form which vanishes outside an ϵ-neighborhood of the zero section.

Choice of a section $\sigma : M \to L$ does not affect the de Rham cohomology class of $\sigma^*(\Phi)$, since any two sections are homotopic. Thus the cohomology class $[\sigma^*(\Phi)]$ equals its pullback via the zero section:

$$[\sigma^*(\Phi)] = \frac{1}{2\pi}[\Omega_{12}] = \frac{1}{2\pi}[F_A] = c_1(L).$$

We now focus on the special case of a complex line bundle L with Hermitian metric over a smooth oriented surface Σ. We say that a smooth section $\sigma : \Sigma \to L$ has a *nondegenerate zero* at $p \in \Sigma$ if

$$\sigma(p) = 0 \text{ and } \sigma(\Sigma) \text{ has transverse intersection with the zero section at } p.$$

Such an intersection has a sign, which is positive if and only if the orientations of the zero section and $\sigma(\Sigma)$ add up to give the orientation of the total space of L. If p is a nondegenerate zero of σ, the *index* of σ at p is

$$\omega(\sigma, p) = \begin{cases} 1, & \text{if the intersection at } p \text{ is positive,} \\ -1, & \text{if the intersection is negative.} \end{cases}$$

Theorem. *If L is a complex unitary line bundle with unitary connection over a smooth compact oriented surface Σ, and $\sigma : \Sigma \to L$ is a section with only nondegenerate zeros, say at $p_1, \ldots, p_k$, then*

$$\sum_{i=1}^{k} \omega(\sigma, p_i) = \frac{1}{2\pi} \int_M F_A = \langle c_1(L), [\Sigma] \rangle.$$

Sketch of Proof: We consider the sections $\sigma_s = s\sigma$ as the real number s approaches ∞. If $\sigma_i = t_i \circ \sigma$,

$$\sigma_s^* \Phi = \frac{1}{2\pi} e^{-s^2(\sigma_1^2 + \sigma_2^2)} \left[\Omega_{12} + 2s^2(d\sigma_1 + \omega_{12}\sigma_2) \wedge (d\sigma_2 - \omega_{12}\sigma_1) \right],$$

an expression which goes to zero as $s \to \infty$ except in very small neighborhoods $V_1, \ldots, V_k$ centered at $p_1, \ldots, p_k$. We can arrange to choose the moving frame so that ω_{12} vanishes at each p_i. Then

$$\lim_{s \to \infty} \int_M \sigma_s^* \Phi = \sum_{i=1}^{k} \lim_{s \to \infty} \int_{V_i} \frac{1}{\pi} e^{-(s\sigma_1)^2 - (s\sigma_2^2)} d(s\sigma_1) \wedge d(s\sigma_2).$$

According to (1.17), the integral over V_i in the sum must approach ± 1, depending upon whether the intersection of $\sigma(\Sigma)$ with the zero section at p_i is positive or negative. Thus

$$\frac{1}{2\pi} \int_\Sigma F_A = \lim_{s \to \infty} \int_\Sigma \sigma_s^*(\Phi) = \sum_{i=1}^{k} \omega(\sigma, p_i),$$

and our sketch of the proof is complete.

For example, suppose that $L = T\Sigma$ the tangent bundle to a surface Σ which possesses a Riemannian metric. A function $f : M \to \mathbb{R}$ is said to be a *Morse function* if the gradient ∇f of f (with respect to the Riemannian metric) has nondegenerate zeros. It is readily checked that the index of this gradient field is one at every local maximum and local minimum, and minus one at every saddle point, so the sum of the indices at the critical points $p_1, \ldots, p_k$ of f is

$$\sum_{i=1}^{k} \omega(\nabla f, p_i) = (\# \text{ of maxima}) - (\# \text{ of saddle points}) + (\# \text{ of minima}).$$

According to the main theorem of Morse theory [28], the last sum on the right is the Euler characteristic $\chi(\Sigma)$ of Σ. Thus the preceding theorem specializes to yield the Gauss-Bonnet formula for a surface:

$$\langle c_1(T\Sigma), [\Sigma] \rangle = \frac{1}{2\pi} \int_\Sigma F_A = \frac{1}{2\pi} \int_\Sigma K dA = \chi(\Sigma),$$

$K : \Sigma \to \mathbb{R}$ being the Gaussian curvature of the surface.

However, the theorem is far more general; when coupled with the notion of intersection number, it also yields a geometric interpretation of the first Chern class of a complex line bundle over a higher-dimensional smooth manifold M. If Σ and Z are two compact submanifolds of complementary dimension in M which intersect transversely at finitely many points $p_1, \ldots, p_k$, we define the *intersection number* of Σ and Z at p_i to be

$$I(\Sigma, Z, p_i) = \begin{cases} 1, & \text{if the orientations of } T_{p_i}\Sigma \oplus T_{p_i}Z \text{ and } T_{p_i}M \text{ agree,} \\ -1, & \text{if these orientations disagree,} \end{cases}$$

and the intersection number of Σ and Z is $\sum_{i=1}^{k} I(\Sigma, Z, p_i)$. (The notion of intersection number is studied in more detail in books on differential topology, such as [21].)

Corollary. *Suppose that L is a complex line bundle with Hermitian metric over a smooth manifold M and $\sigma : M \to L$ is a section which has transverse intersection with the zero section. Let*

$$Z_\sigma = \sigma^{-1}(\text{zero section}).$$

If $i : \Sigma \to M$ is an imbedding of an oriented surface in M which has transverse intersection with Z_σ, then

$$\langle c_1(L), [\Sigma] \rangle = \text{Intersection number of } \Sigma \text{ with } Z_\sigma.$$

Sketch of Proof: The section σ pulls back to a section $i^*\sigma$ of the line bundle i^*L over Σ. The zeros of this pullback correspond exactly to the points of intersection of Σ with Z_σ. Moreover, if p is a point in the intersection $\Sigma \cap Z_\sigma$, $\omega(i^*\sigma, p)$ is just the intersection number of Σ with Z_σ at p.

A similar interpretation is possible for the second Chern class for a quaternionic line bundle over a compact oriented four-manifold M. The Thom form needed for that case is constructed in [26].

1.7 The universal bundle

An alternative approach to the construction of characteristic classes relies on the universal bundle construction.

Suppose that

$$\mathbf{z} = (z_1, z_2, \ldots, z_i, \ldots), \qquad \mathbf{w} = (w_1, w_2, \ldots, w_i, \ldots)$$

are nonzero elements of infinite-dimensional complex Hilbert space $\mathbb{C}^\infty$. We say that $\mathbf{z}$ and $\mathbf{w}$ are equivalent, written $\mathbf{z} \sim \mathbf{w}$, if

$$z_i = \lambda w_i, \qquad \text{for some} \quad \lambda \in \mathbb{C} - \{0\}.$$

Let $[\mathbf{z}] = [z_1, z_2, \ldots, z_i, \ldots]$ denote the equvalence class of $\mathbf{z}$, $P^\infty\mathbb{C}$ the set of equivalence classes. We call $P^\infty\mathbb{C}$ *infinite-dimensional complex projective space*.

It is not difficult to show that $P^\infty\mathbb{C}$ satisfies the definition of infinite-dimensional smooth Hilbert manifold, as described for example in [24]. Indeed, one can contruct a countable atlas of smooth charts

$$\{(U_1, \phi_1), (U_2, \phi_2), \ldots\}$$

by setting

$$U_i = \{[z_1, z_2, \ldots] \in P^\infty\mathbb{C} : z_i \neq 0\},$$

and defining $\phi_i : U_i \to \mathbb{C}^\infty$ by

$$\phi_i([z_1, z_2, \ldots]) = \left(\frac{z_1}{z_i}, \frac{z_2}{z_i}, \ldots, \frac{z_{i-1}}{z_i}, \frac{z_{i+1}}{z_i}, \ldots \right).$$

It is quickly verified that

$$\phi_j \circ (\phi_i)^{-1} : \phi_i(U_i \cap U_j) \to \phi_j(U_i \cap U_j)$$

is smooth, just as in the case of finite-dimensional projective space.

We can regard $P^\infty\mathbb{C}$ as the space of one-dimensional subspaces $V \subset \mathbb{C}^\infty$. There is a *universal bundle* over $P^\infty\mathbb{C}$ whose total space is

$$E_\infty = \{(V, (z_1, z_2, \ldots)) \in P^\infty\mathbb{C} \times \mathbb{C}^\infty : (z_1, z_2, \ldots) \in V\}.$$

(Note that E_∞ minus the zero section is just $\mathbb{C}^\infty - \{0\}$.) The trivialization $\psi_i : \pi^{-1}(U_i) \to U_i \times \mathbb{C}$ over U_i is defined by

$$\psi_i(V, (z_1, z_2, \ldots)) = (V, z_i),$$

and the transition functions corresponding to the covering $\{U_1, U_2, \ldots\}$ are

$$g_{ij} : U_i \cap U_j \to GL(1, \mathbb{C}), \qquad g_{ij} = \frac{z_i}{z_j}.$$

We have just described a special case of a more general construction. Just like infinite-dimensional projective space, the infinite Grassmannian

$$G_m(\mathbb{C}^\infty) = \{m\text{-dimensional complex subspaces of } \mathbb{C}^\infty\}$$

is an infinite-dimensional smooth manifold, and

$$E_\infty = \{(V, v) \in G_m(\mathbb{C}^\infty) \times \mathbb{C}^\infty : v \in V\}$$

is the total space of a smooth vector bundle over $G_m(\mathbb{C}^\infty)$, called the *universal bundle*. Trivializations and transition functions for this bundle are described in standard references, such as [30].

If M is a smooth manifold, we let $[M, G_m(\mathbb{C}^\infty)]$ denote the space of homotopy classes of maps from M to $G_m(\mathbb{C}^\infty)$.

Universal Bundle Theorem. *If M is a smooth manifold, there is a bijection*

$$\Gamma : [M, G_m(\mathbb{C}^\infty)] \to Vect_m^{\mathbb{C}}(M)$$

defined by $\Gamma(F) = F^*E_\infty$.

Sketch of proof: First note that Proposition 2 from §1.3 shows that Γ is well-defined.

To see that Γ is surjective, suppose that (E, π) is a smooth complex vector bundle of rank m over M. We claim that if M has dimension n, it can be covered by $n + 1$ open sets $\{U_0, U_1, \ldots U_n\}$ whose components are contractible. Indeed, let $\{s_i^k : i \in I_k\}$ denote the k-simplices in a simplicial decomposition of M. Each k-simplex s_i^k has a barycenter b_i^k which is a zero-simplex in the first barycentric subdivision. Let U_k be the union of the open stars of all the barycenters b_i^k in the first barycentric subdivision. Then $\{U_0, U_1, \ldots U_n\}$ is a covering of M and each component of U_k is contractible.

By the Corollary at the end of §1.3, we can use this covering as a trivializing covering for the bundle E. Let

$$\psi_k : \pi^{-1}(U_k) \to U_k \times \mathbb{C}^m$$

be the trivialization over U_k, and compose it with projection on the second factor to obtain

$$\eta_k = \pi_2 \circ \psi_k : \pi^{-1}(U_k) \to \mathbb{C}^m.$$

Finally, let $\{\zeta_0, \zeta_1, \ldots, \zeta_n\}$ be a partition of unity subordinate to the open cover and define $\widetilde{F} : E \to \mathbb{C}^{(n+1)m} \subset \mathbb{C}^\infty$ by

$$\widetilde{F}(e) = (\zeta_0(\pi(e))\eta_0(e), \zeta_1(\pi(e))\eta_1(e), \ldots, \zeta_n(\pi(e))\eta_n(e)).$$

Since $\widetilde{F}$ is injective on each fiber, it induces a continuous map $F : M \to G_m(\mathbb{C}^\infty)$ such that $F(p) = \widetilde{F}(E_p)$. It is easily proven that $F^* E_\infty = E$, and surjectivity is established.

Before proving injectivity we need some preliminaries. Let

$$\mathbb{C}_e^\infty = \{(z_1, z_2, z_3, z_4, \ldots) \in \mathbb{C}^\infty : z_1 = 0, z_3 = 0, \ldots\},$$

the set of elements in $\mathbb{C}^\infty$ in which only the even components can be nonzero,

$$\mathbb{C}_o^\infty = \{(z_1, z_2, z_3, z_4, \ldots) \in \mathbb{C}^\infty : z_2 = 0, z_4 = 0, \ldots\},$$

the set of elements in which only the odd components can be nonzero. Define linear maps

$$\widetilde{T}_e : \mathbb{C}^\infty \to \mathbb{C}_e^\infty, \qquad \widetilde{T}_o : \mathbb{C}^\infty \to \mathbb{C}_o^\infty$$

by

$$\widetilde{T}_e(z_1, z_2, z_3, z_4, \ldots) = (0, z_1, 0, z_2, \ldots),$$
$$\widetilde{T}_o(z_1, z_2, z_3, z_4, \ldots) = (z_1, 0, z_2, 0, \ldots).$$

These induce maps

$$T_e : G_m(\mathbb{C}^\infty) \to G_m(\mathbb{C}_e^\infty) \subset G_m(\mathbb{C}^\infty),$$
$$T_o : G_m(\mathbb{C}^\infty) \to G_m(\mathbb{C}_o^\infty) \subset G_m(\mathbb{C}^\infty),$$

which we claim are homotopic to the identity. Indeed, we can define

$$\widetilde{H}_e : \mathbb{C}^\infty \times [0, 1] \to \mathbb{C}^\infty \qquad \text{by} \qquad \widetilde{H}_e(\mathbf{z}, t) = tz + (1 - t)\widetilde{T}_e(z).$$

If $\mathbf{z}_1, \ldots, \mathbf{z}_k$ are linearly independent elements of $\mathbb{C}^\infty$, then so are

$$\widetilde{H}_e(\mathbf{z}_1, t), \quad \ldots, \quad \widetilde{H}_e(\mathbf{z}_k, t)$$

for every choice of $t \in [0, 1]$. Hence $\widetilde{H}_e$ induces a homotopy

$$H_e : G_m(\mathbb{C}^\infty) \times [0, 1] \to G_m(\mathbb{C}^\infty)$$

from T_e to the identity. A similar construction gives a homotopy from T_o to the identity.

To see that Γ is injective, we need to show that if $F, G : M \to G_m(\mathbb{C}^\infty)$ are two smooth maps such that $F^* E_\infty = G^* E_\infty$, then F and G are homotopic. To do this, it suffices to show that $T_e \circ F$ and $T_o \circ G$ are homotopic. But the maps $T_e \circ F$ and $T_o \circ G$ are covered by

$$\widetilde{T}_e \circ \widetilde{F} : E \to \mathbb{C}_e^\infty, \qquad \widetilde{T}_o \circ \widetilde{G} : E \to \mathbb{C}_o^\infty.$$

We can define

$$\widetilde{H} : E \times [0,1] \to \mathbb{C}^\infty \quad \text{by} \quad \widetilde{H}(e,t) = t\, \widetilde{T}_e \circ \widetilde{F}(e,t) + (1-t)\, \widetilde{T}_o \circ \widetilde{G}(e,t).$$

Then $\widetilde{H}|(E_p \times \{t\})$ is a vector space monomorphism for each $(p,t) \in M \times [0,1]$, and hence $\widetilde{H}$ induces a continuous map

$$H : M \times [0,1] \to G_m(\mathbb{C}^\infty), \qquad H(p,t) = \widetilde{H}(E_p \times \{t\}).$$

This map is the desired homotopy from F to G, and injectivity of Γ is established.

We will use the preceding theorem in the case where $m = 1$ to classify complex line bundles. We need a few facts about the homotopy and homology groups of $G_1(\mathbb{C}^\infty) = P^\infty\mathbb{C}$, facts that are worked out in books on algebraic topology. We refer the reader to standard references, such as [37] or [40] for the few topological results that we need.

First of all, the universal bundle restricts to a fiber bundle with fiber $\mathbb{C} - \{0\}$,

$$\mathbb{C} - \{0\} \to (\mathbb{C}^\infty - \{0\}) \to P^\infty\mathbb{C},$$

which yields an exact homotopy sequence (as described in [37], §17)

$$\to \pi_k(\mathbb{C}^\infty - \{0\}) \to \pi_k(P^\infty\mathbb{C}) \to \pi_{k-1}(\mathbb{C} - \{0\}) \to \pi_{k-1}(\mathbb{C}^\infty - \{0\}) \to \tag{1.18}$$

Now $\mathbb{C}^\infty - \{0\}$ is homotopy equivalent to an infinite-dimensional sphere, and just as $\pi_k(S^n) = 0$ for $k < n$, the k-th homotopy group of an infinite-dimensional sphere must vanish for all k. Thus $\pi_k(\mathbb{C}^\infty - \{0\}) = 0$ for all $k > 0$ and hence

$$\pi_k(P^\infty(\mathbb{C})) = \pi_{k-1}(\mathbb{C} - \{0\}) = \begin{cases} \mathbb{Z} & \text{if } k = 2, \\ 0 & \text{otherwise.} \end{cases}$$

In other words, $P^\infty(\mathbb{C})$ is a $K(\mathbb{Z},2)$, an Eilenberg-MacLane space (as described on pages 244-250 of [40]).

On the other hand, this Eilenberg-MacLane space has the remarkable property ([40], page 250) that if M is any smooth manifold,

$$[M, K(\mathbb{Z},2)] = \{\text{homotopy classes of maps } M \to K(\mathbb{Z},2)\} \cong H^2(M;\mathbb{Z}).$$

Combining this with the isomorphism of the Universal Bundle Theorem yields a bijection

$$\text{Vect}_1^{\mathbb{C}}(M) \cong H^2(M;\mathbb{Z}). \tag{1.19}$$

This proves the Classification Theorem for Complex Line Bundles stated in §1.2.

The mapping that realizes the isomorphism (1.19) is (up to sign) just an integral version of the first Chern class. To see this, we could use the theory of differential forms on infinite-dimensional smooth manifolds, together with its corresponding version of de Rham cohomology. Using this theory we could define a connection on the line bundle E_∞ and the first Chern class $c_1(E_\infty) \in H^2(P^\infty(\mathbb{C}); \mathbb{R})$. Naturality would then imply that

$$c_1(E) = F^*(c_1(E_\infty)), \qquad \text{where} \qquad \Gamma(F) = E.$$

Alternatively, we can avoid working with infinite-dimensional manifolds by using N-dimensional projective space $P^N(\mathbb{C})$ as a finite-dimensional approximation to $P^\infty(\mathbb{C})$:

$$P^N(\mathbb{C}) = \{[z_1, z_2, \ldots] \in P^\infty(\mathbb{C}) : z_{N+2} = z_{N+3} = \cdots = 0\}.$$

The argument for the Universal Bundle Theorem shows that for line bundles over M it suffices to take $N = \dim M + 1$.

Let's adopt the second procedure and put two homotopy sequences like (1.18) together to obtain

$$\cdots \;\to\; \pi_2(\mathbb{C}^2 - \{0\}) \;\to\; \pi_2(P^1(\mathbb{C})) \;\to\; \pi_1(\mathbb{C} - \{0\}) \;\to\; \cdots$$
$$\downarrow \qquad\qquad\qquad \downarrow \qquad\qquad\qquad \downarrow$$
$$\cdots \;\to\; \pi_2(\mathbb{C}^{N+1} - \{0\}) \;\to\; \pi_2(P^N(\mathbb{C})) \;\to\; \pi_1(\mathbb{C} - \{0\}) \;\to\; \cdots.$$

Since $\pi_i(\mathbb{C}^2 - \{0\}) \cong \pi_i(\mathbb{C}^{N+1} - \{0\}) \cong 0$ for $i = 1$ or 2, application of the five-lemma shows that the top horizontal arrow in the diagram

$$\pi_2(P^1(\mathbb{C})) \quad \to \quad \pi_2(P^N(\mathbb{C}))$$
$$\downarrow \qquad\qquad\qquad\qquad \downarrow$$
$$H_2(P^1(\mathbb{C}); \mathbb{Z}) \quad \to \quad H_2(P^N(\mathbb{C}); \mathbb{Z})$$

is an isomorphism. The vertical arrows are isomorphisms by the Hurewicz isomorphism theorem ([37], page 79), so the bottom horizontal arrow is also an isomorphism. The universal coefficient theorem then shows that

$$H^2(P^1(\mathbb{C}); \mathbb{Z}) \to H^2(P^N(\mathbb{C}); \mathbb{Z}) \quad \text{and} \quad H^2(P^1(\mathbb{C}); \mathbb{R}) \to H^2(P^N(\mathbb{C}); \mathbb{R}) \tag{1.20}$$

are both isomorphisms.

Let E_∞ denote the universal bundle over $P^N(\mathbb{C})$, a line bundle with total space

$$E_\infty = \{(V, (z_1, \ldots z_{N+1})) \in P^N\mathbb{C} \times \mathbb{C}^N : (z_1, \ldots z_{N+1}) \in V\}. \tag{1.21}$$

We claim that $c_1(E_\infty)$ is the image under the coefficient homomorphism

$$H^2(P^N(\mathbb{C}); \mathbb{Z}) \to H^2(P^N(\mathbb{C}); \mathbb{R})$$

of a generator of $H^2(P^N(\mathbb{C}); \mathbb{Z})$. Because of the isomorphisms (1.20) it suffices to check this for the universal bundle H^{-1} over $P^1(\mathbb{C}) = S^2$. But the Gauss-Bonnet theorem implies that

$$c_1(H^2)[S^2] = c_1(TS^2)[S^2] = \int_{S^2} (1/2\pi) K \, dA = 2.$$

Moreover, it follows from Proposition 4 from §1.5 that if L_1 and L_2 are line bundles,

$$c_1(L_1 \otimes L_2) = c_1(L_1) + c_1(L_2).$$

Hence $c_1(H)[S^2] = 1$ and $c_1(H^{-1})[S^2] = -1$.

Using naturality, we see that under the isomorphism (1.19),

$$F^* E_\infty \mapsto F \in [M, P^\infty(\mathbb{C})] \mapsto F^*(\text{generator of } H^2(P^\infty(\mathbb{C}); \mathbb{Z})).$$

This element of $H^2(M; \mathbb{Z})$ maps to $-c_1(E)$ under the coefficient homomorphism. In particular, the first Chern class of a line bundle E over any smooth manifold M is "quantized"; $c_1(E)$ integrates to an integer over any two-dimensional cycle in M.

This finishes our proof of the Classification Theorem for Complex Line Bundles from §1.2.

There is a real version of the Universal Bundle Theorem which is proven by exactly the same method: If M is a smooth manifold, there is a bijection

$$\Gamma : [M, G_m(\mathbb{R}^\infty)] \to \text{Vect}_m^{\mathbb{R}}(M)$$

defined by $\Gamma(F) = F^* E_\infty$, where E_∞ is the universal bundle defined over the real Grassmannian $G_m(\mathbb{R}^\infty)$.

Just as in the case of the first Chern class, higher degree Chern classes and Pontrjagin classes can be defined by pulling back Chern or Pontrjagin classes of universal bundles over $G_m(\mathbb{C}^\infty)$ or $G_m(\mathbb{R}^\infty)$. Using cohomology with $\mathbb{Z}_2$-coefficients, one can also define Stiefel-Whitney classes. Details of these constructions can be found in [30].

There is also a version of the Universal Bundle Theorem for quaternionic bundles which can be used to prove the Classification Theorem for Quaternionic Line Bundles from §1.2. If $P^\infty(\mathbb{H})$ denotes infinite-dimensional quaternionic projective space, then just as in the complex case there is a bijection

$$\Gamma : [M, P^\infty(\mathbb{H})] \to \text{Vect}_1^{\mathbb{H}}(M).$$

Unfortunately, unlike the complex case, this quaternionic projective space is not an Eilenberg-MacLane space. Nevertheless, a homotopy sequence argument shows that it does satisfy the weaker condition

$$\pi_k(P^\infty(\mathbb{H})) = \pi_{k-1}(\mathbb{H} - \{0\}) = \begin{cases} \mathbb{Z} & \text{if } k = 4, \\ 0 & \text{if } 0 \le k \le 3. \end{cases}$$

This turns out to be sufficient to show that

$$\{\text{homotopy classes of maps } M \to P^\infty(\mathbb{H})\} \cong H^4(M;\mathbb{Z}),$$

when M is a manifold of dimension ≤ 4, thereby establishing the Classification Theorem for Quaternionic Line Bundles.

1.8 Classification of connections

Now that we have classified the complex line bundles over a smooth manifold M, we turn to the problem of classifying the unitary connections in a given complex line bundle L over M.

At first glance, the classification might appear to be trivial. Indeed, we claim that if d_{A_0} is a unitary connection on L, chosen as base point, then any unitary connection on L can be written uniquely in the form

$$d_{A_0} - ia, \qquad \text{where} \qquad a \in \Omega^1(M). \tag{1.22}$$

To establish the claim, we work first in a local trivialization, in terms of which the connection d_{A_0} takes the form

$$(d_{A_0}\sigma)_\alpha = d\sigma_\alpha + \omega_\alpha \sigma_\alpha,$$

where ω_α is a purely imaginary one-form on U_α. The local representative of any other unitary connection in L is

$$(d_A\sigma)_\alpha = d\sigma_\alpha + \omega_\alpha \sigma_\alpha - ia_\alpha \sigma_\alpha,$$

where a_α is a real-valued one-form on U_α. Using the formula (1.8) one quickly verifies that

$$a_\alpha = a_\beta \qquad \text{on the overlap} \quad U_\alpha \cap U_\beta,$$

Thus the a_α's fit together into a real-valued one form a on M and (1.22) holds.

In other words, the space $\mathcal{A}$ of unitary connections on L is an "affine space," isomorphic to the space of one-forms on M. However, we really want to classify connections up to *isomorphism*, also known as *gauge equivalence*.

In the case of $U(1)$-bundles, a *gauge transformation* of L is just a smooth map $g : M \to S^1$, where S^1 is regarded as the complex numbers of length one. A gauge transformation induces a vector bundle isomorphism $g : L \to L$, g acting by scalar multiplication. Conversely, any vector bundle isomorphism of L over M which preserves the Hermitian metric is induced by a gauge transformation.

Let $\mathcal{G}$ denote the a space of gauge transformation, a group under multiplication in S^1. If p_0 is some choice of base point in M, let

$$\mathcal{G}_0 = \{g \in \mathcal{G} : g(p_0) = 1\}.$$

Elements of $\mathcal{G}_0$ are called *based gauge transformations*. We have a direct product of groups,

$$\mathcal{G} = \mathcal{G}_0 \times S^1,$$

where S^1 is the group of constant gauge transformations.

Gauge transformations act on connections by conjugation,

$$(g, d_A) \mapsto g \circ d_A \circ g^{-1} = d_a + g d(g^{-1}).$$

Constant gauge transformations act trivially on the space $\mathcal{A}$, while the group $\mathcal{G}_0$ acts freely. Thus the space of equivalence classes of connections on L is

$$\mathcal{B} = \mathcal{A}/\mathcal{G} = \mathcal{A}/\mathcal{G}_0.$$

It is easy to describe the topology of these spaces. (To be precise, we should complete the spaces $\mathcal{A}$ and $\mathcal{G}$ with respect to suitable norms, such as C^k norms or the Sobolev norms that will be described in more detail in §3.2.) In the case where M is simply connected any based gauge transformation

$$g : M \to S^1, \qquad g(p_0) = 1,$$

has a global logarithm; it can be written in the form

$$g = e^{iu}, \qquad \text{where} \quad u : M \to \mathbb{R}, \quad u(p_0) = 0.$$

Since the space of maps from M to $\mathbb{R}$ is contractible, so is $\mathcal{G}_0$, and hence $\mathcal{G}$ is homotopy equivalent to S^1. In this case, since $\mathcal{A}$ and $\mathcal{G}_0$ are both contractible, so is the space $\mathcal{B}$ of connections.

If M is not simply connected, the Universal Coefficient Theorem implies that $H^1(M; \mathbb{Z})$ is a free abelian group on b_1 generators, where b_1 is the first Betti number of M. Since S^1 is a $K(\mathbb{Z}, 1)$, the theorem on Eilenberg-MacLane spaces cited earlier implies that

$$[M, S^1] = \{ \text{homotopy classes of maps } M \to S^1 \} \cong H^1(M; \mathbb{Z}) \cong \mathbb{Z}^{b_1}.$$

Thus the components of $\mathcal{G}_0$ are in one-to-one correspondence with $\mathbb{Z}^{b_1}$, each component being contractible by an argument similar to that given for the simply connected case. The exact homotopy sequence of the fibration

$$\mathcal{G}_0 \to \mathcal{A} \to \mathcal{B}$$

then shows that each connected component of $\mathcal{B}$ is a $K(G, 1)$, where $G = \mathbb{Z}^{b_1}$, so each component of $\mathcal{B}$ is homotopy equivalent to a torus of rank b_1.

Actually, there is also a far more concrete description of $\mathcal{B}$, which comes from asking the question: If F is an ordinary real-valued two-form on M, when is $\Omega = -iF$ the curvature of a unitary connection in some line bundle over M?

One necessary condition is that F satisfy the Bianchi identity

$$dF = 0.$$

From the previous section, we also know that the de Rham cohomology class of $(1/2\pi)F$ must lie in the image of the coefficient homomorphism

$$H^2(M; \mathbb{Z}) \to H^2(M; \mathbb{R}).$$

We claim that these two conditions are sufficient.

Assume first that M is simply connected. Let L be a complex line bundle over M and let $\mathcal{C}$ denote the space of closed two-forms on M which represent the first Chern class of L. Define

$$\Gamma : \mathcal{B} \to \mathcal{C} \qquad \text{by} \qquad \Gamma([A]) = \frac{1}{2\pi} F_A,$$

where $-iF_A$ is the curvature of A. We claim that Γ is an isomorphism.

To see that Γ is surjective, choose a unitary connection A_0 on L to serve as base point. If F is any element of $\mathcal{C}$, we can write

$$F - F_{A_0} = da,$$

where a is a real-valued one-form on M. Then $d_A = d_{A_0} - ia$ is a unitary connection on L which has curvature $-iF$.

To see that Γ is injective, note that if

$$d_{A_0} - ia_1 \qquad \text{and} \qquad d_{A_0} - ia_2$$

are two unitary connections on L which have the same curvature, then $da_1 = da_2$. Since $H^1(M; \mathbb{R}) = 0$ by hypothesis,

$$a_1 - a_2 = d\phi, \qquad \text{for some } \phi : M \to \mathbb{R} \text{ such that } \phi(p_0) = 0.$$

Then $e^{i\phi} : M \to S^1$ is a gauge transformation such that

$$d_{A_0} - ia_1 = e^{i\phi} \circ (d_{A_0} - ia_2) \circ (e^{i\phi})^{-1},$$

and hence the two connections are gauge equivalent.

It is possible to establish a similar result when M is not simply connected. Suppose, for simplicity, that $H_1(M; \mathbb{Z})$ is a free abelian group of rank b_1 and that $\gamma_1, \ldots, \gamma_{b_1}$ are oriented smooth simple closed curves representing generators for $H_1(M; \mathbb{Z})$, all passing through a given point $p_0 \in M$. Given a unitary connection $d_A \in \mathcal{A}$, parallel translation around γ_i defines an isomorphism

$$\tau_i : L_{p_0} \to L_{p_0}.$$

Since the connection is assumed to be unitary, the isomorphism τ_i is simply a rotation through some angle

$$\tau_i = e^{i\theta_i(A)},$$

which is invariant under gauge transformations. This isomorphism τ_i is called the *holonomy* around γ_i.

Theorem on Classification of Connections. *If $H_1(M; \mathbb{Z})$ is free abelian of rank b_1, the map*

$$\Gamma : \mathcal{B} \to \mathcal{C} \times S^1 \times \cdots \times S^1$$

defined by

$$\Gamma([A]) = \left(\frac{1}{2\pi} F_A, e^{i\theta_1(A)}, \ldots, e^{i\theta_{b_1}(A)} \right)$$

is a bijection.

To see that Γ is injective, suppose that

$$d_{A_0} - ia_1 \qquad \text{and} \qquad d_{A_0} - ia_2$$

are two unitary connections on L which have the same curvature and holonomy. Let $\tilde{M}$ be the universal cover of M, $\tilde{a}_1$ and $\tilde{a}_2$ the pullbacks of a_1 and a_2 to the universal cover. Then $d\tilde{a}_1 = d\tilde{a}_2$ and hence

$$\tilde{a}_1 - \tilde{a}_2 = d\tilde{\phi}, \qquad \text{for some } \tilde{\phi} : M \to \mathbb{R}.$$

Using the fact that the two connections have the same holonomy, one can check that the map $e^{i\tilde{\phi}} : \tilde{M} \to S^1$ descends to a map $e^{i\phi} : M \to S^1$ such that

$$d_{A_0} - ia_1 = e^{i\phi} \circ (d_{A_0} - ia_2) \circ (e^{i\phi})^{-1},$$

and hence the two connections are gauge equivalent. We leave it to the reader to check that Γ is surjective.

This Theorem has an interesting application to electricity and magnetism. Recall that in general relativity (as explained for example in [31]), space-time is a four-dimensional manifold with a pseudo-Riemannian metric of Lorentz signature. It can be proven that any point p in such a manifold lies in a normal coordinate system $(x^0, x^1, x^2, x^3) = (t, x, y, z)$ in terms of which the metric is expressed as

$$ds^2 = -dt^2 + dx^2 + dy^2 + dz^2 + \sum_{i,j=0}^{3} h_{ij} dx^i dx^j,$$

where
$$h_{ij}(p) = 0, \qquad \frac{\partial h_{ij}}{\partial x_k}(p) = 0.$$

What is the right form for Maxwell's equations (the equations of electricity and magnetism) on such a manifold? The classical approach using vector analysis is tied to flat space-time. It turns out that Maxwell's equations can be expressed quite nicely in terms of differential forms, in a way that carries over to curved space-time.

In the case of flat space-time with coordinates (t, x, y, z), we set

$$F = -E_x dt \wedge dx - E_y dt \wedge dy - E_z dt \wedge dz$$

$$+ B_x dy \wedge dz + B_y dz \wedge dx + B_z dx \wedge dy,$$

where $E = (E_x, E_y, E_z)$ and $B = (B_x, B_y, B_z)$ are the electric and magnetic fields. We agree to call the two-form F the *Faraday tensor*. Then two of Maxwell's equations,

$$\mathrm{div}(B) = 0, \qquad \mathrm{curl}(E) + \frac{\partial B}{\partial t} = 0,$$

can be expressed in the simple equation

$$dF = 0.$$

The other two Maxwell equations utilize the "Hodge star," a linear operator $\star : \Lambda^2 T^* M \to \Lambda^2 T^* M$, which interchanges the role of E and B:

$$\star F = B_x dt \wedge dx + B_y dt \wedge dy + B_z dt \wedge dz$$

$$+ E_x dy \wedge dz + E_y dz \wedge dx + E_z dx \wedge dy.$$

The remaining Maxwell equations are (with appropriate choice of units)

$$\text{div}(E) = \rho, \qquad \text{curl}(B) - \frac{\partial E}{\partial t} = J,$$

where ρ is the charge density and

$$J = (J_x, J_y, J_z)$$

is the current density. This pair of equations can be reexpressed as

$$d \star F = \phi$$

where

$$\phi = \rho dx \wedge dy \wedge dz - J_x dt \wedge dy \wedge dz - J_y dt \wedge dz \wedge dx - J_z dt \wedge dx \wedge dy.$$

Now the exterior derivative operator and the Hodge star can be defined on an arbitrary pseudo-Riemannian manifold of Lorentz signature. Thus as described in more detail in [31], Chapter 4, Maxwell's equations

$$dF = 0, \qquad d \star F = \phi,$$

can be formulated in terms of the Faraday tensor on the curved space-times of general relativity.

We can now ask the question: When is $F = F_A$, where $-iF_A$ is the curvature of a unitary connection d_A on some line bundle L over M with Hermitian metric? Note that since F is closed, it represents a cohomology class $[F] \in H^2(M; \mathbb{R})$. The preceding theorem states that $-iF$ is the curvature of a unitary connection precisely when this cohomology class lies in the image of coefficient homomorphism

$$H^2(M; \mathbb{Z}) \to H^2(M; \mathbb{R}).$$

This fact can be interpreted as requiring quantization of magnetic charge. Maintaining a perfect duality between electric and magnetic fields would then require quantization of electric charge as well.

What does one gain by regarding the electromagnetic field as a unitary connection on a line bundle instead of as a two-form? First, a connection gives a family of one-forms, the "electromagnetic gauges," which are useful for solving Maxwell's equations. Second, the connection allows for holonomy around closed curves, which might be detected experimentally (the Bohm-Aharonov effect). Third, regarding the electromagnetic field as a connection suggests fruitful generalizations to other structure groups which may help explain other basic forces (such as weak and strong interactions). Indeed, this is the basis for the standard model for interactions between elementary particles, which is described in [35], for example.

1.9   Hodge theory

The Hodge star, which is used to formulate Maxwell's equations in the case of metrics of Lorentz signature, also plays an important role in four-manifolds with positive-definite Riemannian metrics.

Let p be a point in an oriented four-dimensional Riemannian manifold M and let $V = T_pM$. We can use the Riemannian metric to identify V with the cotangent space T_p^*M. If (e_1, e_2, e_3, e_4) is a positively oriented orthonormal basis for V, then

$$e_1 \wedge e_2, \quad e_1 \wedge e_3, \quad e_1 \wedge e_4, \quad e_2 \wedge e_3, \quad e_2 \wedge e_4, \quad e_3 \wedge e_4$$

forms an orthonormal basis for $\Lambda^2 V$. We define the *Hodge star*

$$\star \Lambda^2 V \to \Lambda^2 V$$

by

$$\star(e_i \wedge e_j) = e_r \wedge e_s,$$

whenever (i, j, r, s) is an even permutation of $(1, 2, 3, 4)$.

More generally, we define the Hodge star

$$\star \Lambda^p V \to \Lambda^{4-p} V$$

by

$$\star 1 = e_1 \wedge e_2 \wedge e_3 \wedge e_4, \qquad \star e_i = e_r \wedge e_s \wedge e_t,$$

where (i, r, s, t) is an even permutation of $(1, 2, 3, 4)$, and so forth. The Hodge star is invariant under the action of the orthogonal group and satisfies the identity $\star^2 = (-1)^p$ on p-forms.

If M is compact, we can use the Hodge star to define a bilinear form

$$(\ , \ ) : \Omega^p(M) \times \Omega^p(M) \to \mathbb{R} \qquad \text{by} \qquad (\omega, \theta) = \int_M \omega \wedge \star \theta.$$

It is easily verified that this bilinear form is symmetric and positive-definite, and hence it is an inner product on $\Omega^p(M)$. We can also use the Hodge star to define the *codifferential*

$$\delta = - \star d\star : \Omega^p(M) \to \Omega^{p-1}(M),$$

which is the *formal adjoint* of the exterior derivative, as one verifies by the calculation,

$$(d\omega, \theta) = \int_M d\omega \wedge \star\theta = \int_M d(\omega \wedge \star\theta) - (-1)^p \int_M \omega \wedge d(\star\theta)$$

1.9. *HODGE THEORY* 41

$$= -\int_M \omega \wedge (\star \star d \star \theta) = \int_M \omega \wedge \star \delta\theta = (\omega, \delta\theta),$$

where $p = \deg \omega$. Finally, we can define the *Hodge Laplacian*

$$\Delta = d\delta + \delta d : \Omega^p(M) \to \Omega^p(M).$$

In the case of $\Omega^0(M)$, the Hodge Laplacian reduces to the standard Laplace operator on functions.

Using the fact that d and δ are formal adjoints of each other, we find that on a compact manifold M without boundary,

$$(\Delta\omega, \omega) = ((d\delta + \delta d)\omega, \omega) = (\delta\omega, \delta\omega) + (d\omega, d\omega) \geq 0.$$

On such a manifold,

$$\Delta\omega = 0 \quad \Leftrightarrow \quad (\Delta\omega, \omega) = 0 \quad \Leftrightarrow \quad d\omega = 0 \quad \text{and} \quad \delta\omega = 0.$$

These last two equations can be thought of as analogs of Maxwell's equations in the case where the charge density and current density vanish.

Definition. A differential p-form ω on a smooth oriented Riemannian manifold M is *harmonic* if $\Delta\omega = 0$. Let $\mathcal{H}^p(M)$ denote the space of harmonic p-forms on M.

Hodge's Theorem. *Every de Rham cohomology class on a compact oriented Riemannian manifold M possesses a unique harmonic representative. Thus*

$$H^p(M; \mathbb{R}) \cong \mathcal{H}^p(M).$$

Moreover, $\mathcal{H}^p(M)$ is finite-dimensional and $\Omega^p(M)$ possesses direct sum decompositions

$$\Omega^p(M) = \mathcal{H}^p(M) \oplus \Delta(\Omega^p(M)) = \mathcal{H}^p(M) \oplus d(\Omega^{p-1}(M)) \oplus \delta(\Omega^{p+1}(M)),$$

which are orthogonal with respect to the inner product (,).

For the proof, we refer the reader to Chapter 6 of [39].

One nice application of Hodge's Theorem is to existence of solutions to Poisson's equation

$$\Delta f = g.$$

If M is connected, then the space $\mathcal{H}^0(M)$ of harmonic zero-forms is just the space of constant functions. Thus Hodge's Theorem implies that if the smooth function $g : M \to \mathbb{R}$ is orthogonal to the constant functions with repect to (,), then $g = \Delta f$ for some smooth function f. Of course, g is orthogonal to the constant functions if and only if its average value is zero,

$$\int_M g \star 1 = 0.$$

A second application of Hodge's Theorem is to topology. Since $\star$ takes harmonic forms to harmonic forms, we see that if M is a compact oriented Riemannian manifold without boundary,

$$\mathcal{H}^p(M) \cong \mathcal{H}^{n-p}(M),$$

and hence one has an isomorphism of de Rham cohomology,

$$H^p(M;\mathbb{R}) \cong H^{n-p}(M;\mathbb{R}).$$

This isomorphism is known as *Poincaré duality* (for cohomology with real coefficients). If b_p denotes the p-th Betti number of M, defined by $b_p = \dim H^p(M;\mathbb{R})$, then Poincaré duality implies that $b_p = b_{n-p}$.

In particular, the Betti numbers of a compact oriented four-manifold are determined by b_0 (which is 1 if M is connected), b_1 (which is 0 if M is simply connected) and b_2. The second Betti number b_2 possesses a further decomposition, which we now describe.

Since $\star^2 = 1$ on two-forms, we can divide $\Lambda^2 V$ into a direct sum,

$$\Lambda^2 V = \Lambda^2_+ V \oplus \Lambda^2_- V,$$

where

$$\Lambda^2_+ V = \{\omega \in \Lambda^2 V : \star\omega = \omega\}, \qquad \Lambda^2_- V = \{\omega \in \Lambda^2 V : \star\omega = -\omega\}.$$

Thus $\Lambda^2_+ V$ is generated by

$$e_1 \wedge e_2 + e_3 \wedge e_4, \quad e_1 \wedge e_3 + e_4 \wedge e_2, \quad e_1 \wedge e_4 + e_2 \wedge e_3,$$

while $\Lambda^2_- V$ is generated by

$$e_1 \wedge e_2 - e_3 \wedge e_4, \quad e_1 \wedge e_3 - e_4 \wedge e_2, \quad e_1 \wedge e_4 - e_2 \wedge e_3.$$

Sections of the bundles whose fibers are $\Lambda^2_+ V$ and $\Lambda^2_- V$ are called *self-dual* and *anti-self-dual* two-forms respectively. If ω is any smooth two-form on M, it can be divided into two orthogonal components,

$$\omega^+ = P_+(\omega) = \frac{1}{2}(\omega + \star\omega) \in \Omega^2_+(M) = \{ \text{ self-dual two-forms on } M \},$$

$$\omega^- = P_-(\omega) = \frac{1}{2}(\omega - \star\omega) \in \Omega_-^2(M) = \{ \text{ anti-self-dual two-forms on } M \ \}.$$

Since $\star$ interchanges the kernels of the operators d and δ, the self-dual and anti-self-dual components of a harmonic two-form are again harmonic. Thus the space of harmonic two-forms on M divides into a direct sum decomposition

$$\mathcal{H}^2(M) \cong \mathcal{H}_+^2(M) \oplus \mathcal{H}_-^2(M),$$

the two summands consisting of self-dual and anti-self-dual harmonic two-forms, respectively. Let

$$b_+ = \dim \mathcal{H}_+^2(M), \qquad b_- = \dim \mathcal{H}_-^2(M).$$

Then $b_+ + b_- = b_2$, the second Betti number of M, while

$$\tau(M) = b_+ - b_-$$

is called the *signature* of M.

On a compact oriented four-manifold, the operator $d^+ = P_+ \circ d :$ $\Omega^1(M) \to \Omega_+^2(M)$ fits into a *fundamental elliptic complex*

$$0 \to \Omega^0(M) \to \Omega^1(M) \to \Omega_+^2(M) \to 0. \tag{1.23}$$

Hodge's Theorem allows us to calculate the cohomology groups of this complex. Indeed, if $\omega \in \Omega_+^2(M)$ is (,)-orthogonal to the image of d^+, then $\delta\omega = 0$. Self-duality then implies that $d\omega = 0$, so ω is harmonic, and $\omega \in \mathcal{H}_+^2(M)$. On the other hand, suppose that $\omega \in \Omega^1(M)$ lies in the kernel of d^+ and is perpendicular to the image of d. Then $\delta\omega = 0$ and $d \star \omega = 0$, and hence

$$(d + \delta)(\omega + \star\omega) = d\omega + \delta \star \omega = d\omega + \star d\omega = 2d^+\omega = 0.$$

Thus $\omega \in \mathcal{H}^1(M)$, and the cohomology groups of the complex (1.23) are just

$$\mathcal{H}^0(M) \cong R, \qquad \mathcal{H}^1(M), \qquad \mathcal{H}_+^2(M).$$

This leads to an immediate proof of the following

Proposition. *Let L be a complex line bundle over the compact oriented Riemannian manifold M. An element $\phi \in \Omega_+^2(M)$ will be the self-dual part of the curvature of some unitary connection on L if and only if ϕ lies in an affine subspace Π of $\Omega_+^2(M)$ of codimension b_+.*

Indeed, we can choose a base connection d_{A_0} on L and let Π be the image of the map

$$a \mapsto F_{A_0}^+ + (da)^+.$$

Then Π is the required affine space of codimension b_+.

From Hodge's Theorem and the classification theorem for connections presented in the previous section, we see that every complex line bundle with Hermitian inner product over a compact oriented simply connected Riemannian four-manifold M has a canonical connection, one with harmonic curvature.

This connection has an important variational characterization. We define the *Yang-Mills functional*

$$\mathcal{Y} : \mathcal{A} \to R \quad\text{by}\quad \mathcal{Y}(d_A) = \int_M F_A \wedge \star F_A.$$

A critical point of this functional is a *Yang-Mills connection* or an *abelian instanton* (said to be abelian because the Lie group $U(1)$ is abelian). Since

$$\mathcal{Y}(d_A - ita) = \int_M (F_A + tda) \wedge \star(F_A + tda)$$

$$= \mathcal{Y}(d_A) + 2t \int_M da \wedge \star F_A + t^2(\cdots)$$

$$= \mathcal{Y}(d_A) - 2t \int_M a \wedge (d \star F_A) + t^2(\cdots),$$

we see that

$$\frac{d}{dt}\mathcal{Y}(d_A - ita)\Big|_{t=0} = -2 \int_M a \wedge (d \star F_A),$$

so the critical points of the Yang-Mills functional are solutions to the Yang-Mills equation

$$d(\star F_A) = 0.$$

Since $dF_A = 0$ by the Bianchi identity, we see that a Yang-Mills connection is just a connection whose curvature form is harmonic.

The Yang-Mills equation can be generalized to quaternionic line bundles, thus leading to a nonabelian gauge theory, which is the foundation for Donaldson's original approach to the geometry and topology of four-manifolds discussed in [14].

Chapter 2

Spin geometry on four-manifolds

2.1 Euclidean geometry and the spin groups

The fact that $\mathbb{R}^2$ possesses a complex multiplication making it into a field
has important applications, leading for example to the theory of Riemann
surfaces. Similarly, the quaternion multiplication on $\mathbb{R}^4$ has important
applications to the geometry of four-manifolds. In four dimensions, quaternions yield a simplification of the theory of spinors (which is presented in
full generality in the highly recommended references [25] and [4]).

We will consider Euclidean four-space to be the space V of quaternions,
complex 2×2 matrices of the form

$$Q = \begin{pmatrix} t + iz & -x + iy \\ x + iy & t - iz \end{pmatrix},$$

where $i = \sqrt{-1}$. Recall that as a real vector space, V is generated by the
four matrices

$$\mathbf{1} = \begin{pmatrix} 1 & 0 \\ 0 & 1 \end{pmatrix}, \quad \mathbf{i} = \begin{pmatrix} 0 & -1 \\ 1 & 0 \end{pmatrix}, \quad \mathbf{j} = \begin{pmatrix} 0 & i \\ i & 0 \end{pmatrix}, \quad \mathbf{k} = \begin{pmatrix} i & 0 \\ 0 & -i \end{pmatrix},$$

the matrix product restricting to minus the cross product on the subspace
spanned by $\mathbf{i}$, $\mathbf{j}$ and $\mathbf{k}$. The determinant of a quaternion Q is given by the
formula

$$\det Q = t^2 + z^2 + x^2 + y^2 = \langle Q, Q \rangle,$$

where $\langle\,,\,\rangle$ denotes the Euclidean dot product. Moreover,

$${}^t\bar{Q}Q = (t^2 + x^2 + y^2 + z^2)I,$$

so the unit sphere in Euclidean four-space can be identified with the special unitary group

$$SU(2) = \{Q \in V : \langle Q, Q \rangle = 1\},$$

while Euclidean four-space itself can be regarded as the set of real multiples of $SU(2)$ matrices. The sphere of unit quaternions $SU(2)$ forms a Lie group under quaternion multiplication, whose Lie algebra—the tangent space to the unit sphere at the identity—is the linear space of "purely imaginary quaternions" spanned by $\mathbf{i}$, $\mathbf{j}$ and $\mathbf{k}$, or equivalently, the linear space of trace-free 2×2 skew-Hermitian matrices.

The four-dimensional spin group is simply the direct product of two copies of the special unitary group,

$$\mathrm{Spin}(4) = SU_+(2) \times SU_-(2);$$

a typical element is (A_+, A_-), where $A_\pm \in SU_\pm(2)$. We have a representation

$$\rho : \mathrm{Spin}(4) \to GL(V) = \{ \text{ isomorphisms from } V \text{ to itself } \}$$

defined by

$$\rho(A_+, A_-)(Q) = A_- Q (A_+)^{-1}.$$

Since A_+ and A_- have determinant one,

$$\langle A_- Q(A_+)^{-1}, A_- Q(A_+)^{-1} \rangle = \det(A_- Q(A_+)^{-1}) = \det Q = \langle Q, Q \rangle,$$

and hence the representation preserves the Euclidean inner product. In other words,

$$\rho : \mathrm{Spin}(4) \to SO(4) \subset GL(V).$$

To get an idea of how this action works, consider the element

$$A = \begin{pmatrix} e^{i\theta} & 0 \\ 0 & e^{-i\theta} \end{pmatrix} \in SU(2).$$

Then

$$\rho(A, I) \begin{pmatrix} t + iz & \natural \\ x + iy & \natural \end{pmatrix} = \begin{pmatrix} t + iz & \natural \\ x + iy & \natural \end{pmatrix} \begin{pmatrix} e^{-i\theta} & 0 \\ 0 & e^{i\theta} \end{pmatrix}$$

$$= \begin{pmatrix} e^{-i\theta}(t + iz) & \natural \\ e^{-i\theta}(x + iy) & \natural \end{pmatrix},$$

a rotation through the angle θ in the same direction in the (t, z)- and (x, y)-planes. On the other hand, one checks that $\rho(I, A)$ rotates the (t, z)- and (x, y)-planes in opposite directions.

More generally, any element $A \in SU(2)$ is conjugate to an element of the form

$$\begin{pmatrix} e^{i\theta} & 0 \\ 0 & e^{-i\theta} \end{pmatrix},$$

and hence there is a positively oriented orthonormal basis (e_1, e_2, e_3, e_4) of V such that $\rho(A, I)$ rotates the planes $e_1 \wedge e_2$ and $e_3 \wedge e_4$ through an angle θ in the same direction, while $\rho(I, A)$ rotates $e_1 \wedge e_2$ and $e_3 \wedge e_4$ in opposite directions.

Since $SO(4)$ is generated by rotations constructed as above, any element of $SO(4)$ lies in the image of ρ and $\rho : \mathrm{Spin}(4) \to SO(4)$ is surjective. The image and range are both compact Lie groups of the same dimension and ρ induces an isomorphism on the level of Lie algebras, so the kernel K of ρ is a finite subgroup and ρ is a smooth covering. Topologically, $\mathrm{Spin}(4)$ is a product $S^3 \times S^3$, hence simply connected, while $\pi_1(SO(4), I) = \mathbb{Z}_2$. The homotopy exact sequence

$$\pi_1(\mathrm{Spin}(4), I) \to \pi_1(SO(4), I) \to K \to 0$$

shows that K is isomorphic to Z_2 and one easily checks that

$$K = \{(I, I), (-I, -I)\}.$$

We can regard $\mathrm{Spin}(4)$ as the space of (4×4)-matrices

$$\begin{pmatrix} A_+ & 0 \\ 0 & A_- \end{pmatrix}, \qquad \text{where} \quad A_\pm \in SU_\pm(2).$$

This Lie group of dimension six is contained in an important Lie group of dimension seven,

$$\mathrm{Spin}(4)^c = \left\{ \begin{pmatrix} \lambda A_+ & 0 \\ 0 & \lambda A_- \end{pmatrix} : A_+ \in SU_+(2),\ A_- \in SU_-(2),\ \lambda \in U(1) \right\}.$$

Moreover, the representation ρ described above extends to a representation

$$\rho^c : \mathrm{Spin}(4)^c \to GL(V),$$

given by the formula

$$\rho^c \begin{pmatrix} \lambda A_+ & 0 \\ 0 & \lambda A_- \end{pmatrix} (Q) = (\lambda A_-) Q (\lambda A_+)^{-1}.$$

We also have a group homomorphism $\pi : \mathrm{Spin}(4)^c \to U(1)$ defined by

$$\pi \begin{pmatrix} \lambda A_+ & 0 \\ 0 & \lambda A_- \end{pmatrix} = \det(\lambda A_+) = \det(\lambda A_-) = \lambda^2.$$

The spin groups Spin(4) and Spin(4)^c fit into exact sequences

$$0 \to Z_2 \to \text{Spin}(4) \to SO(4) \to 0, \tag{2.1}$$

$$0 \to Z_2 \to \text{Spin}(4)^c \to SO(4) \times U(1) \to 0.$$

The key advantage to using the spin groups is that these groups have representations which are more basic that the representations we have described on Euclidean space V itself. Indeed, each factor has its own basic unitary representation. Let W_+ and W_- be two copies of $\mathbb{C}^2$ with its standard hermitian metric $\langle \, , \, \rangle$, one each for $SU_+(2)$ and $SU_-(2)$. Then Spin(4) acts on W_+ and W_- by

$$\rho_+ \begin{pmatrix} A_+ & 0 \\ 0 & A_- \end{pmatrix} (w_+) = A_+ w_+,$$

$$\rho_- \begin{pmatrix} A_+ & 0 \\ 0 & A_- \end{pmatrix} (w_-) = A_- w_-.$$

Similarly, Spin(4)^c acts on W_+ and W_- by

$$\rho_+ \begin{pmatrix} \lambda A_+ & 0 \\ 0 & \lambda A_- \end{pmatrix} (w_+) = \lambda A_+ w_+,$$

$$\rho_- \begin{pmatrix} \lambda A_+ & 0 \\ 0 & \lambda A_- \end{pmatrix} (w_-) = \lambda A_- w_-.$$

These actions preserve the standard Hermitian metrics on W_+ and W_-, and we have an isomorphism of representation spaces

$$V \otimes \mathbb{C} \cong \text{Hom}(W_+, W_-).$$

Since unit-length elements of V are represented by unitary matrices, they act as isometries from W_+ to W_-.

Remark: In physics one is often concerned with four-dimensional space-times of Lorentz signature, and the relevant group is $SL(2, \mathbb{C})$ instead of Spin(4). In this case, a point of flat space-time can be thought of as a 2×2 Hermitian matrix

$$X = \begin{pmatrix} t+z & x-iy \\ x+iy & t-z \end{pmatrix} = tI + x\sigma_x + y\sigma_y + z\sigma_z,$$

where σ_x, σ_y and σ_z are the so-called *Pauli matrices*. Note that

$$\det X = t^2 - x^2 - y^2 - z^2 = -\langle X, X \rangle,$$

where $\langle \, , \, \rangle$ is the usual Lorentz metric of special relativity. An element $A \in SL(2, \mathbb{C})$ acts on Hermitian matrices by $X \mapsto AX^t\bar{A}$, and since $\det(AX^t\bar{A}) = \det X$, $SL(2, \mathbb{C})$ covers the component of the identity in the Lorentz group.

2.2 What is a spin structure?

Suppose that $(M, \langle \ , \ \rangle)$ is an oriented Riemannian manifold of dimension four. The Riemannian metric enables us to reduce the structure group of TM from $GL(4, \mathbb{R})$ to $SO(4)$. Thus we can choose a trivializing cover $\{U_\alpha : \alpha \in A\}$ for TM so that the corresponding transition functions take their values in $SO(4)$:

$$g_{\alpha\beta} : U_\alpha \cap U_\beta \to SO(4) \subset GL(4, \mathbb{R}).$$

Definition 1. A *spin structure* on $(M, \langle , \rangle)$ is given by an open covering $\{U_\alpha : \alpha \in A\}$ of M and a collection of transition functions

$$\tilde{g}_{\alpha\beta} : U_\alpha \cap U_\beta \to \mathrm{Spin}(4)$$

such that $\rho \circ \tilde{g}_{\alpha\beta} = g_{\alpha\beta}$ and the cocycle condition

$$\tilde{g}_{\alpha\beta}\tilde{g}_{\beta\gamma} = \tilde{g}_{\alpha\gamma} \quad \text{on} \quad U_\alpha \cap U_\beta \cap U_\gamma$$

is satisfied.

Manifolds which admit spin structures are called *spin manifolds*. Topologists have found a nice necessary and sufficient condition for existence of a spin structure: M admits a spin structure if and only if $w_2(TM) = 0$, where $w_2(TM)$ represents the second Stiefel-Whitney class of the tangent bundle, an element of $H^2(M; \mathbb{Z}_2)$.

Indeed, the second Stiefel-Whitney class can be defined in terms of Čech cohomology as follows. We say that an open covering $\{U_\alpha : \alpha \in A\}$ of M is a *good cover* (following [8], page 42) if it satisfies the condition,

$$U_{\alpha_1} \cap U_{\alpha_2} \cap \cdots \cap U_{\alpha_k} \quad \text{is either empty or diffeomorphic to } \mathbb{R}^4,$$

for every choice of $(\alpha_1, \ldots, \alpha_k)$. Suppose that we choose a good cover as trivializing cover for TM. Since $U_\alpha \cap U_\beta$ is contractible, each

$$g_{\alpha\beta} : U_\alpha \cap U_\beta \to SO(4) \quad \text{can be lifted to} \quad \tilde{g}_{\alpha\beta} : U_\alpha \cap U_\beta \to \mathrm{Spin}(4).$$

The problem is that the cocycle condition may not hold. However, by exactness of (2.1),

$$\eta_{\alpha\beta\gamma} = \tilde{g}_{\alpha\beta}\tilde{g}_{\beta\gamma}\tilde{g}_{\gamma\alpha} : U_\alpha \cap U_\beta \cap U_\gamma \longrightarrow \{\pm 1\} = \mathbb{Z}_2.$$

It is readily verified that

$$\eta_{\alpha\beta\gamma} = \eta_{\beta\gamma\alpha} = \eta_{\beta\alpha\gamma} = \eta_{\beta\alpha\gamma}^{-1},$$

the last equality holding because elements of $\mathbb{Z}_2$ are their own inverses, and

$$\eta_{\beta\gamma\delta}\eta_{\alpha\gamma\delta}^{-1}\eta_{\alpha\beta\delta}\eta_{\alpha\beta\gamma}^{-1} = 1.$$

Thus in the language of Čech cohomology,

$$\{\eta_{\alpha\beta\gamma} : (\alpha,\beta,\gamma) \in A \times A \times A\}$$

is a Čech cocycle which, since the cover is good, represents a cohomology class in $H^2(M;\mathbb{Z}_2)$. That cohomology class is called the *second Stiefel-Whitney class* of the tangent bundle and is denoted by $w_2(TM)$.

If M has a spin structure, we can choose the $\tilde{g}_{\alpha\beta}$'s to satisfy the cocycle condition, so the second Stiefel-Whitney class must be zero. Conversely, if the second Stiefel-Whitney class is zero, the Čech theory implies that $\{\eta_{\alpha\beta\gamma}\}$ is a coboundary, which means that there exist constant maps

$$\eta_{\alpha\beta} : U_\alpha \cap U_\beta \longrightarrow Z_2 \quad \text{such that} \quad \eta_{\alpha\beta}\eta_{\beta\gamma}\eta_{\gamma\alpha} = \eta_{\alpha\beta\gamma}.$$

Then $\{\eta_{\alpha\beta}\tilde{g}_{\alpha\beta}\}$ are transition functions defining a spin structure on M.

Recall that once we have transition functions satifying the cocycle condition, we get a corresponding vector bundle. Thus if we have a spin structure on $(M, \langle\ ,\ \rangle)$ defined by

$$\tilde{g}_{\alpha\beta} : U_\alpha \cap U_\beta \to \text{Spin}(4)$$

the transition functions

$$\rho_+ \circ \tilde{g}_{\alpha\beta} : U_\alpha \cap U_\beta \to SU_+(2), \qquad \rho_- \circ \tilde{g}_{\alpha\beta} : U_\alpha \cap U_\beta \to SU_-(2)$$

determine complex vector bundles of rank two over M, which we denote by W_+ and W_-. Moreover,

$$TM \otimes \mathbb{C} \cong \text{Hom}(W_+, W_-),$$

so a spin structure allows us to represent the complexified tangent bundle of M in terms of two more basic complex vector bundles. These bundles W_+ and W_- can be regarded as quaternionic line bundles over M. More generally, if L is a complex line bundle over M, we can also write

$$TM \otimes \mathbb{C} \cong \text{Hom}(W_+ \otimes L, W_- \otimes L),$$

because the transition functions for L cancel out.

Definition 2. A *spin^c structure* on $(M, \langle,\rangle)$ is given by an open covering $\{U_\alpha : \alpha \in A\}$ and a collection of transition functions

$$\tilde{g}_{\alpha\beta} : U_\alpha \cap U_\beta \to \text{Spin}(4)^c$$

such that $\rho \circ \tilde{g}_{\alpha\beta} = g_{\alpha\beta}$ and the cocycle condition is satisfied.

If $(M, \langle , \rangle)$ has a spin structure defined by the transition functions

$$\tilde{g}_{\alpha\beta} : U_\alpha \cap U_\beta \to \mathrm{Spin}(4)$$

and L is a complex line bundle over M with Hermitian metric and transition functions

$$h_{\alpha\beta} : U_\alpha \cap U_\beta \to U(1),$$

we can define a spin^c structure on M by taking the transition functions to be

$$h_{\alpha\beta} \tilde{g}_{\alpha\beta} : U_\alpha \cap U_\beta \to \mathrm{Spin}(4)^c.$$

In fact, on a simply connected spin manifold, isomorphism classes of spin^c structures on a given Riemannian manifold are in one-to-one correspondence with complex line bundles L over M. However, spin^c structures also exist on manifolds which do not have genuine spin structures:

Theorem. *Every compact oriented four-manifold possesses a spin^c structure.*

Sketch of proof: The key fact we need is that the second Stiefel-Whitney class is the $\mathbb{Z}_2$-reduction of an integral cohomology class. For simply connected four-manifolds (the case we need for subsequent applications), this fact follows from the long exact sequence corresponding to the sequence of coefficient groups

$$1 \longrightarrow \mathbb{Z} \longrightarrow \mathbb{Z} \longrightarrow \mathbb{Z}_2 \longrightarrow 1$$

and the fact that $H^3(M; \mathbb{Z}) = 0$ by Poincaré duality. (For the general case, the reader may wish to consult Appendix D of [25].)

If $\{U_\alpha : \alpha \in A\}$ is a good trivializing cover for TM, we now know that there is a Čech cocycle $\eta_{\alpha\beta\gamma} : U_\alpha \cap U_\beta \cap U_\gamma \to \{\pm 1\}$ representing $w_2(TM)$ which lifts to an integral cocycle

$$\tilde{\eta}_{\alpha\beta\gamma} : U_\alpha \cap U_\beta \cap U_\gamma \longrightarrow \mathbb{Z} \quad \text{so that} \quad \exp(\pi i \tilde{\eta}_{\alpha\beta\gamma}) = \eta_{\alpha\beta\gamma}.$$

Here $\tilde{\eta}_{\alpha\beta\gamma}$ changes sign under an odd permutation of the indices and the cocycle condition is

$$\tilde{\eta}_{\beta\gamma\delta} - \tilde{\eta}_{\alpha\gamma\delta} + \tilde{\eta}_{\alpha\beta\delta} - \tilde{\eta}_{\alpha\beta\gamma} = 0. \tag{2.2}$$

Let $\{\psi_\alpha : \alpha \in A\}$ be a partition of unity subordinate to $\{U_\alpha : \alpha \in A\}$ and define

$$f_{\beta\gamma} : U_\beta \cap U_\gamma \longrightarrow R \quad \text{by} \quad f_{\beta\gamma} = \sum_\alpha \psi_\alpha \tilde{\eta}_{\alpha\beta\gamma}.$$

Then a straightforward calculation using (2.2) shows that

$$f_{\alpha\beta} + f_{\beta\gamma} + f_{\gamma\alpha} = \sum_\delta \psi_\delta \tilde{\eta}_{\delta\alpha\beta} + \sum_\delta \psi_\delta \tilde{\eta}_{\delta\beta\gamma} + \sum_\delta \psi_\delta \tilde{\eta}_{\delta\gamma\alpha}$$

$$= \sum_\delta \psi_\delta \tilde{\eta}_{\alpha\beta\gamma} = \tilde{\eta}_{\alpha\beta\gamma}.$$

Thus if we set

$$h_{\alpha\beta} = \exp(\pi i f_{\alpha\beta}) : U_\alpha \cap U_\beta \longrightarrow U(1),$$

we find that the $h_{\alpha\beta}$'s fail to satisfy the cocycle condition for a line bundle with exactly the same discrepancy as the $\tilde{g}_{\alpha\beta}$'s:

$$h_{\alpha\beta} h_{\beta\gamma} h_{\gamma\alpha} = \eta_{\alpha\beta\gamma}$$

It follows that the maps

$$h_{\alpha\beta}^{-1} \tilde{g}_{\alpha\beta} : U_\alpha \cap U_\beta \to \mathrm{Spin}(4)^c$$

do satisfy the cocycle condition and therefore define a spin^c structure on M, finishing our sketch of the proof.

Given a spin^c structure defined by the transition functions

$$\tilde{g}_{\alpha\beta} : U_\alpha \cap U_\beta \to \mathrm{Spin}(4)^c,$$

the transition functions

$$\pi \circ \tilde{g}_{\alpha\beta} : U_\alpha \cap U_\beta \to U(1)$$

determine a complex line bundle. We denote this bundle by L^2 to remind ourselves of the fact that it is the square of the line bundle L used in the construction of a spin^c structure on a spin manifold. It follows from the proof of the above theorem that $w_2(TM)$ is the reduction mod 2 of $c_1(L^2)$.

Similarly, the representations ρ_+^c and ρ_-^c yield $U(2)$-bundles $W_+ \otimes L$ and $W_- \otimes L$. Note that the bundles $W_+ \otimes L$ and $W_- \otimes L$ exist as genuine vector bundles even though the factors W_+, W_- and L do not unless M is spin. In these notes, we will call W_+, W_- and L *virtual vector bundles*.

Just as in the case of spin structures, we see that a spin^c structure allows us to represent the complexified tangent bundle in terms of two more basic bundles,

$$TM \otimes \mathbb{C} \cong \mathrm{Hom}(W_+ \otimes L, W_- \otimes L).$$

Sections of $W_+ \otimes L$ and $W_- \otimes L$ are called *spinor fields* of positive or negative chirality, respectively.

2.3 Almost complex and spin^c structures

Suppose now that M is a complex manifold of complex dimension two. Thus M has complex coordinate systems

$$(z_1, z_2) = (x_1 + iy_1, x_2 + iy_2)$$

which are related holomorphically on overlaps. We can define a complex multiplication

$$J : TM \to TM \quad \text{by} \quad J\left(\frac{\partial}{\partial x_i}\right) = \frac{\partial}{\partial y_i}, \quad J\left(\frac{\partial}{\partial y_i}\right) = -\frac{\partial}{\partial x_i}.$$

This is a vector bundle endomorphism which satisfies the identity $J^2 = -I$.

More generally, an *almost complex structure* on a manifold M is simply a vector bundle endomorphism

$$J : TM \to TM \quad \text{such that} \quad J^2 = -I.$$

An almost complex structure on a four-manifold enables us to construct trivializations of the tangent bundle TM so that the corresponding transition functions take their values in $GL(2, \mathbb{C}) \subset GL(4, \mathbb{R})$. This enables us to regard TM as a complex vector bundle.

A Riemannian metric $\langle, \rangle$ on an almost complex manifold M is said to be *Hermitian* if it satisfies the condition

$$\langle Jv, Jw \rangle = \langle v, w \rangle,$$

for all $v, w \in T_pM$. Such a metric can be constructed using a partition of unity. The Hermitian metric reduces the structure group further from $GL(2, \mathbb{C})$ to $U(2) \subset O(4)$. We claim that this allow us to construct a canonical spin^c structure on M.

Indeed, we have a canonical embedding $j : U(2) \to \mathrm{Spin}(4)^c$ defined by

$$j(A) = \begin{pmatrix} 1 & 0 & 0 & 0 \\ 0 & \det A & 0 & 0 \\ 0 & 0 & & \\ 0 & 0 & & A \end{pmatrix}$$

such that

$$\rho^c\left(j(A), \begin{pmatrix} t + iz & -x + iy \\ x + iy & t - iz \end{pmatrix}\right) = A\begin{pmatrix} t + iz & \sharp \\ x + iy & \sharp \end{pmatrix}.$$

This is the usual unitary action on $\mathbb{C}^2$ with coordinates $(t + iz, x + iy)$ and hence

$$\rho^c \circ j : U(2) \to SO(4)$$

is the usual inclusion.

Thus a $U(2)$-structure

$$\{g_{\alpha\beta} : U_\alpha \cap U_\beta \to U(2) : \alpha, \beta \in A\}$$

on M determines a corresponding spin^c structure

$$\{\tilde{g}_{\alpha\beta} = j \circ g_{\alpha\beta} : U_\alpha \cap U_\beta \to \mathrm{Spin}(4)^c : \alpha, \beta \in A\}.$$

As described in the preceding section, this allows us to define the complex vector bundles $W_+ \otimes L$ and $W_- \otimes L$.

Of course, when M has a *bona fide* spin structure, these vector bundles are just the tensor products of the spin bundles W_+ and W_- with a complex line bundle L.

In the general case, L is not well-defined, but L^2 is. It has the transition functions

$$\pi \circ \tilde{g}_{\alpha\beta} : U_\alpha \cap U_\beta \longrightarrow U(1)$$

where $\pi : \mathrm{Spin}(4)^c \to U(1)$ is the "determinant map"

$$\pi \begin{pmatrix} \lambda A_+ & 0 \\ 0 & \lambda A_- \end{pmatrix} = \det(\lambda A_-) = \lambda^2.$$

The line bundle L^2 will be called the *anticanonical bundle* of the $U(2)$-structure. Note that $W_- \otimes L$ can be identified with the holomorphic tangent bundle of the complex manifold M and the anticanonical bundle L^2 is isomorphic to the second exterior power of the holomorphic tangent bundle.

2.4   Clifford algebras

Instead of using the tangent bundle TM as the foundation for tensor algebra on a Riemannian manifold, we can use the more basic bundles W_+ and W_- or $W_+ \otimes L$ and $W_- \otimes L$, where L is a line bundle with Hermitian metric.

Let's start on the vector space level. An element of Euclidean space V can be regarded as a complex linear homomorphism from W_+ to W_- and represented by a quaternion Q. It can be extended to a skew-Hermitian endomorphism of $W = W_+ \oplus W_-$ which is represented by the 2×2 quaternion matrix

$$\theta(Q) = \begin{pmatrix} 0 & -{}^t\bar{Q} \\ Q & 0 \end{pmatrix}.$$

We can extend θ to a complex linear map

$$\theta : V \otimes \mathbb{C} \to \mathrm{End}(W).$$

Note that $\text{End}(W)$ is a sixteen-dimensional algebra over the complex numbers with composition (or matrix multiplication) as the algebra multiplication. If $Q \in V$,

$$(\theta(Q))^2 = \begin{pmatrix} -^t\bar{Q}Q & 0 \\ 0 & -Q^t\bar{Q} \end{pmatrix} = (-\det Q)I = -\langle Q, Q \rangle. \qquad (2.3)$$

Because of this formula, we can regard $\text{End}(W)$ as the *Clifford algebra* of $(V \otimes \mathbb{C}, \langle \, , \, \rangle)$; matrix multiplication in this algebra is referred to as *Clifford multiplication*.

There is another way that is commonly used to obtain the Clifford algebra of $(V \otimes \mathbb{C}, \langle \, , \, \rangle)$. One simply constructs complex 4×4 matrices e_1, e_2, e_3, e_4 which satisfy the identities

$$e_i \cdot e_j + e_j \cdot e_i = -2\delta_{ij} = \begin{cases} -2 & \text{if } i = j, \\ 0 & \text{if } i \neq j. \end{cases}$$

These matrices correspond to the image under θ of an orthonormal basis of V. Although any such choice would work, it is often convenient to have an explicit one in mind. We take

$$e_1 = \begin{pmatrix} 0 & 0 & -1 & 0 \\ 0 & 0 & 0 & -1 \\ 1 & 0 & 0 & 0 \\ 0 & 1 & 0 & 0 \end{pmatrix}, \qquad e_2 = \begin{pmatrix} 0 & 0 & i & 0 \\ 0 & 0 & 0 & -i \\ i & 0 & 0 & 0 \\ 0 & -i & 0 & 0 \end{pmatrix},$$

$$e_3 = \begin{pmatrix} 0 & 0 & 0 & -1 \\ 0 & 0 & 1 & 0 \\ 0 & -1 & 0 & 0 \\ 1 & 0 & 0 & 0 \end{pmatrix}, \qquad e_4 = \begin{pmatrix} 0 & 0 & 0 & i \\ 0 & 0 & i & 0 \\ 0 & i & 0 & 0 \\ i & 0 & 0 & 0 \end{pmatrix}.$$

As a complex vector space $\text{End}(W)$ has a basis consisting of the matrices

$$I, \quad e_i, \quad e_i e_j \ \text{ for } i < j, \quad e_i e_j e_k \ \text{ for } i < j < k, \quad e_1 e_2 e_3 e_4. \qquad (2.4)$$

It follows immediately from (2.3) that

$$e_i e_j = -e_j e_i, \qquad \text{when} \qquad i \neq j.$$

Thus we can identify the complexified second exterior power $\Lambda^2 V \otimes \mathbb{C}$ with the complex subspace of $\text{End}(W)$ generated by the products $e_i \cdot e_j$. Similarly, the third exterior power of V sits inside $\text{End}(W)$. Indeed, the basis (2.4) corresponds to a direct sum decomposition

$$\text{End}(W) = \Lambda^0 V \otimes \mathbb{C} \oplus \Lambda^1 V \otimes \mathbb{C} \oplus \Lambda^2 V \otimes \mathbb{C} \oplus \Lambda^3 V \otimes \mathbb{C} \oplus \Lambda^4 V \otimes \mathbb{C}. \qquad (2.5)$$

We can therefore regard the Clifford algebra as just the complexified exterior algebra with a new multiplication, Clifford product instead of wedge product. In fact, the Clifford product of e_i with an element $\omega \in \sum_{k=0}^{4} \Lambda^k V$ can be defined in terms of the wedge product and the interior product

$$\iota(e_i) : \Lambda^k V \to \Lambda^{k-1} V, \qquad \langle \iota(e_i)\omega, \theta \rangle = \langle \omega, e_i \wedge \theta \rangle$$

by means of the formula

$$e_i \cdot \omega = e_i \wedge \omega - \iota(e_i)\omega. \tag{2.6}$$

As in §1.8, $\Lambda^2 V$ has a direct sum decomposition into self-dual and anti-self-dual parts,

$$\Lambda^2 V = \Lambda_+^2 V \oplus \Lambda_-^2 V,$$

the self-dual part being generated by

$$e_1 \wedge e_2 + e_3 \wedge e_4, \qquad e_1 \wedge e_3 + e_4 \wedge e_2, \qquad e_1 \wedge e_4 + e_2 \wedge e_3.$$

The corresponding elements in the Clifford algebra are

$$e_1 \cdot e_2 + e_3 \cdot e_4 = \begin{pmatrix} -2i & 0 & 0 & 0 \\ 0 & 2i & 0 & 0 \\ 0 & 0 & 0 & 0 \\ 0 & 0 & 0 & 0 \end{pmatrix},$$

$$e_1 \cdot e_3 + e_4 \cdot e_2 = \begin{pmatrix} 0 & 2 & 0 & 0 \\ -2 & 0 & 0 & 0 \\ 0 & 0 & 0 & 0 \\ 0 & 0 & 0 & 0 \end{pmatrix},$$

$$e_1 \cdot e_4 + e_2 \cdot e_3 = \begin{pmatrix} 0 & -2i & 0 & 0 \\ -2i & 0 & 0 & 0 \\ 0 & 0 & 0 & 0 \\ 0 & 0 & 0 & 0 \end{pmatrix}.$$

Thus we see that $\Lambda_+^2 V$ is just the space of trace-free skew-Hermitian endomorphisms of W_+, which is the Lie algebra of $SU_+(2)$. Similarly, $\Lambda_-^2 V$ is the space of trace-free skew-Hermitian endomorphisms of W_-.

If we represent an element $\psi \in W_+$ in terms of its components as

$$\psi = \begin{pmatrix} \psi_1 \\ \psi_2 \end{pmatrix},$$

then if $\langle \cdot, \cdot \rangle$ denotes the standard Hermitian inner product on W, a direct calculation yields

$$\langle \psi, e_1 e_2 \psi \rangle = \langle \psi, e_3 e_4 \psi \rangle = \begin{pmatrix} \psi_1 & \psi_2 \end{pmatrix} \begin{pmatrix} -i & 0 \\ 0 & i \end{pmatrix} \begin{pmatrix} \bar{\psi}_1 \\ \bar{\psi}_2 \end{pmatrix} = -i(|\psi_1|^2 - |\psi_2|^2),$$

$$\langle \psi, e_1 e_3 \psi \rangle = \langle \psi, e_4 e_2 \psi \rangle = (\psi_1 \quad \psi_2) \begin{pmatrix} 0 & 1 \\ -1 & 0 \end{pmatrix} \begin{pmatrix} \bar{\psi}_1 \\ \bar{\psi}_2 \end{pmatrix}$$

$$= -(\bar{\psi}_1 \psi_2 - \psi_1 \bar{\psi}_2) = -2i \operatorname{Im}(\bar{\psi}_1 \psi_2),$$

$$\langle \psi, e_1 e_4 \psi \rangle = \langle \psi, e_2 e_3 \psi \rangle = (\psi_1 \quad \psi_2) \begin{pmatrix} 0 & -i \\ -i & 0 \end{pmatrix} \begin{pmatrix} \bar{\psi}_1 \\ \bar{\psi}_2 \end{pmatrix}$$

$$= -i(\bar{\psi}_1 \psi_2 + \psi_1 \bar{\psi}_2) = -2i \operatorname{Re}(\bar{\psi}_1 \psi_2).$$

We can define a quadratic map $\sigma : W_+ \to \Lambda_+^2 V$ in one of two ways: First, we can identify $\Lambda_+^2 V$ with 2×2 trace-free skew-Hermitian matrices and set

$$\sigma(\psi) = 2i \text{ Trace-free part of } \begin{pmatrix} \bar{\psi}_1 \\ \bar{\psi}_2 \end{pmatrix} (\psi_1 \quad \psi_2)$$

$$= i \begin{pmatrix} |\psi_1|^2 - |\psi_2|^2 & 2\bar{\psi}_1 \psi_2 \\ 2\bar{\psi}_2 \psi_1 & |\psi_2|^2 - |\psi_1|^2 \end{pmatrix}.$$

Equivalently, we can identify $\Lambda_+^2 V$ with trace-free skew-Hermitian matrices lying in the upper left 2×2 block, and set

$$\sigma(\psi) = -\frac{i}{2} \sum_{i<j} \langle \psi, e_i \cdot e_j \cdot \psi \rangle e_i \cdot e_j. \tag{2.7}$$

Since our convention is that $e_i \cdot e_j$ should have length one, we can readily verify that

$$|\sigma(\psi)|^2 = \frac{1}{2}(|\psi_1|^2 - |\psi_2|^2)^2 + 2|\psi_1 \psi_2|^2 = \frac{1}{2}|\psi|^4, \quad \text{or} \quad |\sigma(\psi)| = \frac{1}{\sqrt{2}}|\psi|^2.$$

Note that σ forgets the phase of ψ: $\sigma(e^{i\theta}\psi) = \sigma(\psi)$.

The groups Spin(4) and Spin^c(4) act on the linear space End(W) by conjugation; these actions are denoted by Ad and Ad^c and are called the adjoint representations. If $T \in \text{End}(W)$,

$$\text{Ad} \begin{pmatrix} A_+ & 0 \\ 0 & A_- \end{pmatrix} (T) = \begin{pmatrix} A_+ & 0 \\ 0 & A_- \end{pmatrix} T \begin{pmatrix} A_+^{-1} & 0 \\ 0 & A_-^{-1} \end{pmatrix},$$

while

$$\text{Ad}^c \begin{pmatrix} \lambda A_+ & 0 \\ 0 & \lambda A_- \end{pmatrix} (T) = \begin{pmatrix} \lambda A_+ & 0 \\ 0 & \lambda A_- \end{pmatrix} T \begin{pmatrix} (\lambda A_+)^{-1} & 0 \\ 0 & (\lambda A_-)^{-1} \end{pmatrix}.$$

If $(M, \langle \, , \, \rangle)$ has a spin or spin^c structure, we can use these representations to construct a sixteen-dimensional complex vector bundle over M, a bundle of Clifford algebras. For simplicity, we denote this bundle also by $\mathrm{End}(W)$. Note that this bundle of Clifford algebras is defined even when M is not a spin manifold and W_+ and W_- are not defined, because the adjoint representations factor through $SO(4)$.

Since the actions of $\mathrm{Spin}(4)$ and $\mathrm{Spin}^c(4)$ preserve the direct sum decomposition (2.5), the bundle $\mathrm{End}(W)$ divides into a direct sum decomposition

$$\mathrm{End}(W) = \sum_{k=1}^{4} \Lambda^k TM \otimes \mathbb{C}, \quad \Lambda^2 TM \otimes \mathbb{C} = \Lambda_+^2 TM \otimes \mathbb{C} \oplus \Lambda_-^2 TM \otimes \mathbb{C}.$$

Moreover, since the vector space quadratic map $\sigma : W_+ \to \Lambda_+^2 V$ is equivariant with respect to the group action, it extends to a quadratic map on the vector bundle level $\sigma : W_+ \otimes L \to \Lambda_+^2 TM$.

The trace of the Hermitian matrix

$$\begin{pmatrix} \bar{\psi}_1 \\ \bar{\psi}_2 \end{pmatrix} (\psi_1 \quad \psi_2)$$

is the length of the spinor field ψ. The trace-free part of this matrix is $\sigma(\psi)$, which can be thought of as a field of infinitesimal rotations, rotating two planes $e_1 \wedge e_2$ and $\star(e_1 \wedge e_2)$ through the same angle in the same direction. The quadratic map σ gives a geometric interpretation to spinor fields: a spinor field of positive chirality can be regarded as the square root of a self-dual two-form, together with a choice of phase.

2.5   The spin connection

From Riemannian geometry we know that if $(M, \langle\, ,\, \rangle)$ is a Riemannian manifold, its tangent bundle a canonical connection, the Levi-Civita connection. We claim that when W_+ and W_- are defined, they inherit canonical connections from the Levi-Civita connection.

There are two steps to the construction of these connections. First we construct a connection on the bundle $\mathrm{End}(W)$. This is easy, because the Levi-Civita connection on TM induces a connection on $\Lambda^k TM$ for each k and hence a connection on

$$\mathrm{End}(W) = \sum_{k=1}^{4} \Lambda^k TM \otimes \mathbb{C}.$$

However, we want a connection on W. A connection d_A on W is called a *Spin(4)-connection* if it can be expressed in terms of each local trivialization as

$$(d_A\sigma)_\alpha = d\sigma_\alpha + \phi_\alpha\sigma_\alpha,$$

where ϕ_α is a one-form with values in the Lie algebra of $\mathrm{Spin}(4) = SU_+(2) \times SU_-(2)$. Now the Lie algebra of $\mathrm{Spin}(4)$ is $\Lambda^2 TM$ which is generated by $e_i \cdot e_j$ for $i < j$. Thus the condition that d_A is a $\mathrm{Spin}(4)$-connection is simply

$$\phi_\alpha = \sum_{i<j} \phi_{\alpha ij} e_i \cdot e_j, \qquad \phi_{\alpha ij} = -\phi_{\alpha ji},$$

where $\phi_{\alpha ij}$ are ordinary real-valued one-forms.

Given a $\mathrm{Spin}(4)$-connection d_A on W, there is a unique connection (also to be denoted by d_A) on $\mathrm{End}(W)$ which satisfies the Leibniz rule:

$$d_A(\omega\sigma) = (d_A\omega)\sigma + \omega d_A\sigma, \qquad \text{for} \quad \omega \in \Gamma(\mathrm{End}(W)), \sigma \in \Gamma(W). \quad (2.8)$$

The following theorem allows us to go the other direction:

Theorem 1. *Suppose that M is a four-dimensional oriented Riemannian manifold with a spin structure. Then there is a unique $\mathrm{Spin}(4)$-connection on W which induces the Levi-Civita connection on $\mathrm{End}(W)$.*

Proof: In fact, we claim that there is a one-to-one correspondence between $\mathrm{Spin}(4)$-connections on W and $SO(4)$-connections on TM. To see this, let $\tilde\psi$ be a trivialization of W over an open set $U \subset M$. The trivialization of W determines a trivialization of $\mathrm{End}(W)$ as well as trivializations of subbundles of $\mathrm{End}(W)$ corresponding to linear subspaces of the Clifford algebra which are left fixed by the action of $\mathrm{Spin}(4)$. In particular, $\tilde\psi$ induces a trivialization ψ of TM over U.

Let $(\epsilon_1, \epsilon_2, \epsilon_3, \epsilon_4)$ be the orthonormal sections of $W|U$ defined by

$$\tilde\psi \circ \epsilon_1(p) = \left(p, \begin{pmatrix} 1 \\ 0 \\ 0 \\ 0 \end{pmatrix} \right), \quad \ldots, \quad \tilde\psi \circ \epsilon_4(p) = \left(p, \begin{pmatrix} 0 \\ 0 \\ 0 \\ 1 \end{pmatrix} \right).$$

Let (e_1, e_2, e_3, e_4) be the orthonormal sections of $TM|U$ defined by

$$\psi \circ e_1(p) = \left(p, \begin{pmatrix} 0 & 0 & -1 & 0 \\ 0 & 0 & 0 & -1 \\ 1 & 0 & 0 & 0 \\ 0 & 1 & 0 & 0 \end{pmatrix} \right), \quad \ldots,$$

$$\psi \circ e_4(p) = \left(p, \begin{pmatrix} 0 & 0 & 0 & i \\ 0 & 0 & i & 0 \\ 0 & i & 0 & 0 \\ i & 0 & 0 & 0 \end{pmatrix} \right).$$

Then

$$e_i \cdot \epsilon_\lambda = \sum_{\mu=1}^{4} c_{i\lambda}^\mu \epsilon_\mu,$$

where the $c_{i\lambda}^\mu$'s are constants.

Suppose now that d_A is a Spin(4)-connection on W, given over U by

$$d + \sum_{i,j=1}^{4} \phi_{ij} e_i \cdot e_j, \qquad \phi_{ij} = -\phi_{ji}.$$

Since the components of ϵ_λ and $e_k \cdot \epsilon_\lambda$ are constant,

$$d_A \epsilon_\lambda = \sum_{i,j=1}^{4} \phi_{ij} e_i e_j \epsilon_\lambda, \qquad d_A(e_k \epsilon_\lambda) = \sum_{i,j=1}^{4} \phi_{ij} e_i e_j e_k \epsilon_\lambda.$$

Hence it follows from condition (2.8) that

$$\sum_{i,j=1}^{4} \phi_{ij} e_i e_j e_k \epsilon_\lambda = (d_A e_k) \epsilon_\lambda + e_k \sum_{i,j=1}^{4} \phi_{ij} e_i e_j \epsilon_\lambda,$$

or equivalently,

$$(d_A e_k) \epsilon_\lambda = \sum_{i,j=1}^{4} \phi_{ij} (e_i e_j e_k - e_k e_i e_j) \epsilon_\lambda.$$

The only terms surviving in the sum on the right are those in which $i \neq j$ and $k = i$ or $k = j$. A short calculation shows that

$$d_A e_k = -4 \sum_{i=1}^{4} \phi_{ij} e_i.$$

It follows directly from this formula the the connection d_A on $\text{End}(W)$ preserves TM and is an orthogonal connection.

Conversely, given an orthogonal connection

$$d_A e_k = \sum_{i=1}^{4} \omega_{ik} e_i$$

on TM, we can define a corresponding Spin(4)- connection on W by setting

$$\phi_{ij} = -\frac{1}{4} \omega_{ij}.$$

This is the unique Spin(4)-connection on W which induces the $SO(4)$-connection. In terms of the local trivialization, the connection is simply

$$d - \frac{1}{4} \sum_{i,j=1}^{4} \omega_{ij} e_i \cdot e_j. \tag{2.9}$$

What is the curvature

$$\Omega = \left(d - \frac{1}{4} \sum_{i,j=1}^{4} \omega_{ij} e_i \cdot e_j \right)^2$$

of the spin connection? It is not difficult to answer this question if we remember that in the local expression the e_i's are constant elements of $\text{End}(W)$. A short calculation shows that

$$\Omega = d \left(-\frac{1}{4} \sum_{i,j=1}^{4} \omega_{ij} e_i \cdot e_j \right) + \left(\frac{1}{4} \sum_{i,j=1}^{4} \omega_{ij} e_i \cdot e_j \right)^2 = -\frac{1}{4} \sum_{i,j=1}^{4} \Omega_{ij} e_i \cdot e_j,$$

where

$$\Omega_{ij} = d\omega_{ij} + \sum_{k=1}^{4} \omega_{ik} \wedge \omega_{kj}.$$

We can set

$$R_{ijkl} = \Omega_{ij}(e_k, e_l), \qquad \text{so that} \qquad \Omega_{ij} = \sum_{k,l=1}^{4} R_{ijkl} \theta_k \wedge \theta_l,$$

where $(\theta_1, \theta_2, \theta_3, \theta_4)$ is the coframe dual to (e_1, e_2, e_3, e_4). The R_{ijkl}'s are the components of the *Riemann-Christoffel curvature tensor*. This tensor is studied extensively in Riemannian geometry texts, where it is shown that the components satisfy the curvature symmetries

$$R_{ijkl} = -R_{jikl} = R_{ijlk} = R_{klij}, \qquad R_{ijkl} + R_{iklj} + R_{iljk} = 0.$$

An important invariant of the Riemannian manifold M is the *scalar curvature* $s : M \to \mathbb{R}$, given by the formula

$$s = \sum_{i,j=1}^{4} R_{ijij}.$$

We claim that

$$\sum_{i,j=1}^{4} e_i e_j \Omega(e_i, e_j) = \frac{s}{2}. \tag{2.10}$$

Indeed, since $R_{klij} = R_{ijkl}$,

$$\sum_{i,j=1}^{4} e_i e_j \Omega(e_i, e_j) = -\frac{1}{4} \sum_{i,j,k,l=1}^{4} e_i e_j e_k e_l \Omega_{kl}(e_i, e_j)$$

$$= -\frac{1}{4} \sum_{i,j,k,l=1}^{4} e_i e_j e_k e_l R_{ijkl}.$$

Now observe that if i, j, k and l are distinct,

$$e_i e_j e_k e_l R_{ijkl} + e_i e_k e_l e_j R_{iklj} + e_i e_l e_j e_k R_{iljk}$$

$$= e_i e_j e_k e_l (R_{ijkl} + R_{iklj} + R_{iljk}) = 0,$$

by the last of the curvature symmetries. Similarly, if i, j and l are distinct,

$$e_i e_j e_i e_l R_{ijil} + e_i e_l e_i e_j R_{ilij} = e_i e_j e_i e_l (R_{ijil} - R_{ilij}) = 0.$$

Hence the only terms surviving in the sum are those containing R_{ijij} or R_{ijji}, and

$$\sum_{i,j=1}^{4} e_i e_j \Omega(e_i, e_j) = -\frac{1}{4} \sum_{i,j=1}^{4} e_i e_j e_i e_j R_{ijij} - \frac{1}{4} \sum_{i,j=1}^{4} e_i e_j e_j e_i R_{ijji} = \frac{s}{2}.$$

Given a unitary connection d_A on a complex line bundle L over a spin manifold M, we can define a connection on the bundle $W \otimes L$ by taking the tensor product of this connection with the connection given by Theorem 1. This connection will also be denoted by d_A. It is a Spin(4)^c-connection, which means that it can be expressed in terms of each local trivialization as

$$(d_A \sigma)_\alpha = d\sigma_\alpha + \phi_\alpha \sigma_\alpha,$$

where ϕ_α is a one-form with values in the Lie algebra of Spin(4)^c. In fact,

$$\phi_\alpha = -iaI - \frac{1}{4} \sum_{i,j=1}^{4} \omega_{ij} e_i \cdot e_j,$$

a being an ordinary real-valued one-form. Moreover, $da = F_A$, where $-iF_A$ is the curvature of the original connection on L. The curvature of the new connection on $W \otimes L$ is

$$\Omega_A = -iF_A I + \Omega = -iF_A I - \frac{1}{4} \sum_{i,j=1}^{4} \Omega_{ij} e_i \cdot e_j, \qquad (2.11)$$

a two-form with values in $\mathrm{End}(W)$. In this formula, the Ω_{ij}'s are the curvature forms for the Levi-Civita connection on M.

If M is not a spin manifold, we can still construct bundles $W \otimes L$ for various choices of line bundles L^2. A $\mathrm{Spin}(4)^c$-connection on $W \otimes L$ determines a connection on the Clifford algebra bundle $\mathrm{End}(W)$ and also on the determinant line bundle $\Lambda^2(W_+ \otimes L) = L^2$. When we denote the connection on $W \otimes L$ by d_A, we will denote the induced connection on L^2 by d_{2A}.

Theorem 2. *Suppose that M is a four-dimensional oriented Riemannian manifold with a spin^c structure having determinant line bundle L^2. Given a connection d_{2A} on L^2, there is a unique $\mathrm{Spin}(4)^c$-connection on $W \otimes L$ which induces the Levi-Civita connection on $\mathrm{End}(W)$ and the connection d_{2A} on L^2.*

The proof is similar to that of Theorem 1. The curvature of this connection is still given by (2.11), in agreement with the case of spin manifolds.

2.6 The Dirac operator

Let M be a four-dimensional Riemannian manifold with a spin^c structure and a $\mathrm{Spin}(4)^c$-connection d_A on the spin bundle $W \otimes L$.

Definition. The *Dirac operator* $D_A : \Gamma(W \otimes L) \to \Gamma(W \otimes L)$ with coefficients in the line bundle L is defined by

$$D_A(\psi) = \sum_{i=1}^{4} e_i \cdot d_A \psi(e_i) = \sum_{i=1}^{4} e_i \cdot \nabla_{e_i}^A \psi. \qquad (2.12)$$

In the case where M is four-dimensional Euclidean space with the global Euclidean coordinates (x_1, x_2, x_3, x_4) the bundle W is trivial—$W = M \times \mathbb{C}^4$. If, in addition, L is the trivial line bundle, the Dirac operator is just

$$D_A \psi = \sum_{i=1}^{4} e_i \frac{\partial \psi}{\partial x_i},$$

the e_i's being constant matrices such that

$$e_i \cdot e_j + e_j \cdot e_i = -2\delta_{ij} = \begin{cases} -2 & \text{for } i = j, \\ 0 & \text{for } i \neq j. \end{cases}$$

Thus we find that

$$D_A \circ D_A(\psi) = -\sum_{i=1}^{4} \frac{\partial^2 \psi}{\partial x_i^2}.$$

In this case, the Dirac operator is (up to sign) a square root of the usual Euclidean Laplacian.

The notion of Dirac operator with coefficients in a line bundle can be extended further to coefficients in a general vector bundle. This is a powerful extension which includes many of the familiar first order elliptic operators of differential geometry.

Indeed, if M has a spin structure, and E is a complex vector bundle over M with Hermitian metric and unitary connection, we have an induced Hermitian metric and a unitary connection on $W \otimes E$. If M has only a spin^c structure with spin bundle $W \otimes L$, a Hermitian metric and unitary connection in a complex vector bundle $E \otimes L^{-1}$ will induce a Hermitian metric and unitary structure in $W \otimes E$. In either case we can construct a *Dirac operator* $D_A : \Gamma(W \otimes E) \to \Gamma(W \otimes E)$ with coefficients in E by a straightforward extension of (2.12):

$$D_A(\psi \otimes \sigma) = \sum_{i=1}^{4} e_i \cdot d_A \psi(e_i \otimes \sigma),$$

in which d_A now denotes the connection on $W \otimes E$.

For example, if we take the bundle of coefficient E to be W itself, we obtain in this way a familiar operator from Hodge theory:

Proposition. *The Dirac operator with coefficients in W is*

$$D_W = d + \delta : \sum_{k=0}^{4} \Lambda^k TM \otimes \mathbb{C} \to \sum_{k=0}^{4} \Lambda^k TM \otimes \mathbb{C}.$$

Sketch of proof: First one shows that if ∇ denotes the Levi-Civita connection on the exterior algebra $\sum_{k=0}^{4} \Lambda^k TM$, then the exterior derivative d is given by the formula

$$d\omega = \sum_{i=1}^{4} e_i \wedge \nabla_{e_i} \omega,$$

while the codifferential δ is given by

$$\delta\omega = -\sum_{i=1}^{4} \iota(e_i)\nabla_{e_i}\omega,$$

where $\iota(e_i)$ denotes the interior product. The claimed formula can then be derived from (2.6).

The main result of this section, Weitzenböck's formula, gives a simple relationship between a Dirac operator and the "vector bundle Laplacian" $\Delta^A : \Gamma(W \otimes E) \to \Gamma(W \otimes E)$ defined by the formula

$$\Delta^A\psi = -\sum_{i=1}^{4}[\nabla_{e_i}^A \circ \nabla_{e_i}^A\psi - \nabla_{\nabla_{e_i}e_i}^A\psi], \tag{2.13}$$

where (e_1, e_2, e_3, e_4) is a moving orthonormal frame. (The term $\nabla_{\nabla_{e_i}e_i}^A\psi$ is needed to make $\Delta^A\psi$ independent of the choice of frame.)

Theorem (Weitzenböck's formula). *The square of the Dirac operator with coefficients in a line bundle bundle L is related to the Laplace operator Δ^A by the formula*

$$D_A^2\psi = \Delta^A\psi + \frac{s}{4}\psi - \sum_{i<j} F_A(e_i, e_j)(ie_i \cdot e_j \cdot \psi).$$

Here s is the scalar curvature of M and F_A is the curvature of the connection in L.

Proof: Choose a moving orthonormal frame on a neighborhood of p in M so that $\nabla_{e_i}e_j(p) = 0$. (Recall that the Spin(4)^c-connection ∇^A acts as the Levi-Civita connection ∇ on vector fields.) Then at p,

$$D_A^2\psi = \left(\sum_{i=1}^{4} e_i\nabla_{e_i}^A\right)\left(\sum_{j=1}^{4} e_j\nabla_{e_j}^A\right)\psi = \sum_{i,j=1}^{4} e_ie_j\nabla_{e_i}^A\nabla_{e_j}^A\psi$$

$$= -\sum_{i=1}^{4}\nabla_{e_i}^A\nabla_{e_i}^A\psi + \frac{1}{2}\sum_{i,j=1}^{4} e_ie_j[\nabla_{e_i}^A\nabla_{e_j}^A - \nabla_{e_j}^A\nabla_{e_i}^A]\psi$$

$$= \Delta^A\psi + \frac{1}{2}\sum_{i,j=1}^{4} e_ie_j(d_A^2\psi)(e_i, e_j) = \Delta^A\psi + \frac{1}{2}\sum_{i,j=1}^{4} e_ie_j\Omega_A(e_i, e_j)\cdot\psi,$$

where Ω_A is the curvature of the Spin(4)^c-connection. Substituting (2.11) into this expression and utilizing (2.10) yields

$$D_A^2\psi = \Delta^A\psi - \frac{1}{2}\sum_{i,j=1}^{4} F_A(e_i, e_j)ie_i\cdot e_j\cdot\psi - \frac{1}{8}\sum_{i,j,k,l=1}^{4}\Omega_{kl}(e_i, e_j)e_i\cdot e_j\cdot e_k\cdot e_l\cdot\psi$$

$$= \Delta^A\psi - \sum_{i<j}F_A(e_i, e_j)ie_i\cdot e_j\cdot\psi + \frac{s}{4}\psi,$$

which proves the theorem.

In addition to being almost the square root of a Laplacian, the Dirac operator has another key property—self-adjointness:

Proposition. *The Dirac operator is "formally self-adjoint":*

$$\int_M \langle D_A(\psi), \eta\rangle dV = \int_M \langle\psi, D_A(\eta)\rangle dV \tag{2.14}$$

Proof: (Compare [25], page 114.) As in the preceding proof, we choose a moving orthonormal frame on a neighborhood of p in M so that $\nabla_{e_i}e_j(p) = 0$. Then

$$\langle D_A(\psi), \eta\rangle(p) = \sum_{i=1}^{4}\langle e_i\cdot\nabla_{e_i}^A\psi, \eta\rangle(p) = -\sum_{i=1}^{4}\langle\nabla_{e_i}^A\psi, e_i\cdot\eta\rangle(p)$$

$$= -\sum_{i=1}^{4}\left[e_i\langle\psi, e_i\cdot\eta\rangle(p) - \langle\psi, \nabla_{e_i}^A(e_i\cdot\eta)\rangle(p)\right]$$

$$= -\sum_{i=1}^{4}e_i\langle\psi, e_i\cdot\eta\rangle(p) + \langle\psi, D_A(\eta)\rangle(p),$$

because $e_i\cdot$ is skew-Hermitian and ∇^A is a unitary connection. If we define a one-form b on M by $\langle b, e_i\rangle = \langle\psi, e_i\cdot\eta\rangle$, we can reexpress the result of this calculation as

$$\langle D_A(\psi), \eta\rangle(p) - \langle\psi, D_A(\eta)\rangle(p) = \delta b. \tag{2.15}$$

Integration over M now yields equation (2.14).

Of course the vector bundle Laplacian Δ^A is also formally self-adjoint; this follows directly from the easily-proven integral formula

$$\int_M \langle\psi, \Delta^A\psi\rangle dV = \int_M |\nabla^A\psi|^2 dV. \tag{2.16}$$

2.7 The Atiyah-Singer Index Theorem

In many respects, the simplest Dirac operator with coefficients is the operator $d + \delta$ from Hodge theory, whose square is the Hodge Laplacian $\Delta = d\delta + \delta d$. Since

$$(d + \delta)\omega = 0 \qquad \Leftrightarrow \qquad d\omega = 0 \ \text{ and } \ \delta\omega = 0 \qquad \Leftrightarrow \qquad \Delta\omega = 0,$$

we see that the kernel of $d + \delta$ is the finite-dimensional space of harmonic forms. We can set

$$\Omega^+ = \Omega^0(M) \oplus \Omega^2(M) \oplus \Omega^4(M), \qquad \Omega^- = \Omega^1(M) \oplus \Omega^3(M)$$

and divide the operator $d + \delta$ into two pieces,

$$(d + \delta)^+ : \Omega^+ \to \Omega^-, \qquad (d + \delta)^- : \Omega^- \to \Omega^+.$$

we define the *index* of $(d + \delta)^+$ to be

$$\text{index of } (d + \delta)^+ = \dim(\text{Ker}(d + \delta)^+) - \dim(\text{Ker}(d + \delta)^-).$$

But

$$\text{Ker}((d + \delta)^+) = \mathcal{H}^0(M) \oplus \mathcal{H}^2(M) \oplus \mathcal{H}^4(M),$$

$$\text{Ker}((d + \delta)^-) = \mathcal{H}^1(M) \oplus \mathcal{H}^3(M),$$

so

$$\text{index of } (d + \delta)^+ = b_0 + b_2 + b_4 - (b_1 + b_3) = \chi(M),$$

the Euler characteristic of M.

In a similar spirit, the Atiyah-Singer Index Theorem gives a formula for the index of any first order elliptic linear differential operator in terms of topological data. Of most importance to us is the Dirac operator D_A with coefficients in a line bundle.

First note that the Dirac operator D_A divides into two pieces,

$$D_A^+ : \Gamma(W_+ \otimes L) \to \Gamma(W_- \otimes L), \qquad D_A^- : \Gamma(W_- \otimes L) \to \Gamma(W_+ \otimes L),$$

which are formal adjoints of each other by (2.14). Just as in Hodge theory, the theory of elliptic operators implies that the kernels of the operators D_A^+ and D_A^- are finite-dimensional complex vector spaces. We define the *index* of D_A^+ to be

$$\text{index of } D_A^+ = \dim(\text{Ker}(D_A^+)) - \dim(\text{Ker}(D_A^-)).$$

Atiyah-Singer Index Theorem (for the Dirac operator with coefficients in a line bundle). *If D_A is a Dirac operator with coeffficients in a line bundle L on a compact oriented four-manifold M, then*

$$\text{index of } D_A^+ = -\frac{1}{8}\tau(M) + \frac{1}{2}\int_M c_1(L)^2, \qquad (2.17)$$

where $\tau(M) = b_+ - b_-$, the signature of M.

This theorem is a consequence of a more general theorem to be stated later in this section. It has many striking applications. In particular, if we take L to be the trivial line bundle over a spin manifold, we can derive:

Rochlin's Theorem. *The signature of a compact oriented smooth spin manifold of dimension four satisfies the condition*

$$\tau(M) \equiv 0, \qquad mod \quad 16.$$

Proof: If D^+ is the Dirac operator with coefficients in the trivial line bundle,

$$\text{index of } D^+ = -\frac{\tau(M)}{8}.$$

This shows immediately that the signature of M is divisible by 8.

But there are also endomorphisms $J : W_\pm \to W_\pm$ defined by Clifford multiplication with

$$e_1 \cdot e_3 = \begin{pmatrix} 0 & 1 & 0 & 0 \\ -1 & 0 & 0 & 0 \\ 0 & 0 & 0 & -1 \\ 0 & 0 & 1 & 0 \end{pmatrix}.$$

If ψ is a harmonic section of W_+, so is $J\psi$, a section which is perpendicular to ψ. In fact, the kernel of D^+ is quaternionic, and we can construct an orthonormal basis of $\text{Ker}(D^+)$ of the form

$$\psi_1 J\psi_1, \ldots, \psi_k, J\psi_k.$$

The same argument applies to $\text{Ker}(D^-)$, so $\text{Ker}(D^+)$ and $\text{Ker}(D^-)$ are both even-dimensional. Thus the index of D^+ is divisible by two, and the signature of M is divisible by 16.

Another striking application of the Atiyah-Singer Index Theorem is to the problem of relating curvature to topology of Riemannian manifolds:

Theorem of Lichnerowicz. *If M is a compact oriented spin manifold of dimension four with a Riemannian metric of positive scalar curvature, then the signature of M is zero.*

Proof: We use the Weitzenböck formula for the Dirac operator D in the case of the trivial line bundle with the trivial connection. If ψ is a harmonic spinor field $(D\psi = 0)$, then

$$0 = \int_M \left[|\nabla_A \psi|^2 + \frac{s}{4}|\psi|^2 \right] dv,$$

and hence positive scalar curvature implies that $\psi = 0$. This implies that

$$-\frac{\tau(M)}{8} = (\text{index of } D^+) = 0.$$

This theorem shows that many compact four-manifolds do not admit Riemannian metrics with positive scalar curvature.

Coefficients in a general vector bundle: The Atiyah-Singer Index Theorem with coefficients in a line bundle is a special case of a more general theorem, which gives the index of a Dirac operator

$$D_A^+ : \Gamma(W_+ \otimes E) \to \Gamma(W_- \otimes E)$$

with coefficients in a general vector bundle or virtual vector bundle E. To state the more general theorem, we need the Chern character discussed in §1.5 and the $\hat{A}$-polynomial in the Pontrjagin classes.

It follows from our earlier discussion of the Chern character and (1.15) that

$$\text{ch}(E) = \dim E + c_1(E) + \frac{1}{2}(c_1(E))^2 - c_2(E) + \cdots, \qquad (2.18)$$

the dots denoting terms of degree > 4. We can apply these formulae to the quaternionic line bundles W_+ and W_- which have vanishing first Chern classes:

$$\text{ch}(W_+) = 2 - c_2(W_+), \qquad \text{ch}(W_-) = 2 - c_2(W_-).$$

Since W_+ is isomorphic to its conjugate or dual bundle,

$$TM \otimes \mathbb{C} = \text{Hom}(W_+, W_-) \cong W_+ \otimes W_-,$$

and hence

$$\text{ch}(TM \otimes \mathbb{C}) = \text{ch}(W_+)\text{ch}(W_-) = 4 - 2c_2(W_+) - 2c_2(W_-).$$

Thus according to the definition of Pontrjagin class,

$$p_1(TM) = -c_2(TM \otimes \mathbb{C}) = -2c_2(W_+) - 2c_2(W_-) = -2c_2(W). \quad (2.19)$$

It follows from the product formula for the Chern character that if E is any $SU(2)$-bundle, then

$$(2 - c_2(E) + \cdots)^2 = 4 - c_2(E \otimes E) + \cdots,$$

which implies that

$$c_2(E) = \frac{1}{4}c_2(E \otimes E).$$

If M is not spin, we can use this formula to define $c_2(W_+)$ and $c_2(W_-)$, since the vector bundles $W_+ \otimes W_+$ and $W_- \otimes W_-$ always exist as *bona fide* vector bundles. Similarly, we set $c_1(L) = (1/2)c_1(L^2)$, when L^2 exists as a complex line bundle, but L does not. Thus we can use the product formula for Chern characters to define $c_1(E)$ and $c_2(E)$ when E is a virtual vector bundle.

The $\hat{A}$-polynomial is defined by a power series

$$\hat{A}(TM) = 1 - \frac{1}{24}p_1(TM) + \cdots \qquad \in H^*(M; \mathbb{R}),$$

where p_1 is the first Pontrjagin class and the dots indicate terms in higher order Pontrjagin classes, which once again will vanish on a manifold of dimension ≤ 4.

Atiyah-Singer Index Theorem (over four-manifolds). *If D_A is a Dirac operator with coefficients in a virtual vector bundle E over a compact oriented four-manifold M, then*

$$index \ of \ D_A^+ = \hat{A}(TM)ch(E)[M]$$

$$= \int_M [-(1/24)(\dim E)p_1(TM) + (1/2)(c_1(E))^2 - c_2(E)].$$

For the proof of this theorem, one of the monuments of twentieth century mathematics, we refer the reader to [6] or [25]. This theorem gives a formula for the indices of virtually all the geometrically interesting elliptic operators on a closed four-manifold.

How does this theorem imply the Atiyah-Singer Theorem we previously stated for Dirac operators with coefficients in line bundles? To answer this question, we need to consider first the special case where $E = W$, so that

$$W \otimes E = \sum_{k=1}^{4} \Lambda^k TM \otimes \mathbb{C} \qquad \text{and} \qquad D_A = (d + \delta).$$

In this case,

$$\mathrm{Ker}(D_A) = \mathrm{Ker}(d + \delta) = \{ \text{ complex-valued harmonic forms } \}.$$

This is the operator considered at the beginning of this section, but now we divide up the operator differently—this time we take

$$D_A^+ : \Gamma(W_+ \otimes W) \to \Gamma(W_- \otimes W).$$

It is easiest to understand the new decomposition in the case where M is simply connected, so $b_1 = b_3 = 0$. In this case, the harmonic forms consist of the constant functions, constant multiples of the volume form $\star 1$, and elements of

$$\mathcal{H}_+^2(M) \quad \text{and} \quad \mathcal{H}_-^2(M),$$

which are sections of $W_+ \otimes W_+$ and $W_- \otimes W_-$, respectively. Since $1 + \star 1$ and $1 - \star 1$ are sections of $W_+ \otimes W_+$ and $W_- \otimes W_-$ respectively, we see that

$$\text{the index of } D_A^+ = b_+ - b_- = \tau(M),$$

the signature of M. With a little more work, one can show that this formula also holds in the case where $b_1 = b_3 \neq 0$, and hence by the Atiyah-Singer Index Theorem and (2.19),

$$\tau(M) = \int_M [-(1/24)(\dim W) p_1(TM) - c_2(W)]$$

$$= \int_M [-(1/6) p_1(TM) + (1/2) p_1(TM)] = \int_M (1/3) p_1(TM).$$

We have sketched the proof of:

Hirzebruch Signature Theorem (for four-manifolds). *The signature of a compact oriented smooth four-manifold M is given by the formula*

$$\tau(M) = \frac{1}{3} \int_M p_1(TM).$$

The Signature Theorem allows us to eliminate the term containing $p_1(TM)$ in the general Atiyah-Singer Index Theorem, thereby obtaining (2.17) in the case of coefficients in a line bundle.

We conclude this chapter by deriving the characteristic class invariants for the bundles $W_\pm \otimes L$. From (2.19) we conclude that

$$c_2(W_+)[M] + c_2(W_-)[M] = -\frac{1}{2} \int_M p_1(TM) = -\frac{3}{2} \tau(M).$$

On the other hand, we can take Dirac operators D_{W_+} and D_{W_-} with coefficients in W_+ and W_- respectively, and verify that

the index of $D_{W_+}^+$ $-$ the index of $D_{W_-}^+$ $= 2b_0 - 2b_1 + b_2 = \chi(M)$,

the Euler characteristic of M. Hence by the Atiyah-Singer Theorem,

$$-c_2(W_+)[M] + c_2(W_-)[M] = \chi(M).$$

We can solve for the Chern classes of W_+ and W_-, obtaining

$$c_2(W_+)[M] = -\frac{3}{4}\tau(M) - \frac{1}{2}\chi(M), \qquad c_2(W_-)[M] = -\frac{3}{4}\tau(M) + \frac{1}{2}\chi(M).$$

By the classification theorem for quaternionic line bundles, these numbers completely determine the spinor bundles W_+ and W_-, when they exist.

Applying the product formula for Chern characters to $W_+ \otimes L$ gives

$$(2 - c_2(W_+) + \cdots)(1 + c_1(L) + \frac{1}{2}(c_1(L))^2 + \cdots)$$

$$= 2 + c_1(W_+ \otimes L) + \frac{1}{2}[c_1(W_+ \otimes L)]^2 - c_2(W_+ \otimes L) + \cdots.$$

Thus

$$c_1(W_+ \otimes L) = 2c_1(L) = c_1(L^2)$$

and finally,

$$c_2(W_+ \otimes L)[M] = -\frac{3}{4}\tau(M) - \frac{1}{2}\chi(M) + \frac{1}{4}c_1(L^2)^2[M], \qquad (2.20)$$

$$c_2(W_- \otimes L)[M] = -\frac{3}{4}\tau(M) + \frac{1}{2}\chi(M) + \frac{1}{4}c_1(L^2)^2[M]. \qquad (2.21)$$

Chapter 3

Global analysis of the Seiberg-Witten equations

3.1 The Seiberg-Witten equations

As we have seen, the theory of linear partial differential equations—in particular, the Atiyah-Singer index theorem— yields topological invariants which have striking geometric applications. We now investigate more refined invariants constructed with nonlinear partial differential equations, invariants that are not available from the linear theory.

Here is the procedure we might try to follow. Choose a nonlinear PDE. Show that its space of solutions is a compact finite-dimensional smooth manifold which lies in some ambient "configuration space." The solution space will in general depend upon certain choices, such as the choice of a Riemannian metric, but the "cobordism class" of the solution space may be independent of the choices. This cobordism class may become a new topological invariant. Since it is defined in terms of a PDE (which requires a smooth structure), it may be possible for the invariant to distinguish between smooth structures on the underlying topological manifold associated to M.

In broad outline, this is in fact the way Donaldson's original theory works. The moduli space of anti-self-dual connections in an $SU(2)$-bundle over a compact oriented four-manifold is generically a smooth manifold. However, this moduli space is not usually compact, and much effort is expended towards finding a suitable compactification.

Gromov's theory of pseudoholomorphic curves [27] is a second case in which the above procedure can be carried through, with compactification once again providing difficulties. A body of techniques has been developed for studying these and related geometrical theories. Perhaps the Seiberg-Witten invariants, which we will proceed to describe, provide the simplest context for the working out of these techniques.

Let $(M, \langle \ , \ \rangle)$ be an oriented four-dimensional Riemannian manifold with a spin^c structure and corresponding positive spinor bundle $W_+ \otimes L$. We seek pairs (d_{2A}, ψ), where d_{2A} is a connection on the line bundle L^2 and ψ is a section of $W_+ \otimes L$ such that

$$D_A^+ \psi = 0, \qquad F_A^+(e_i, e_j) = -\frac{i}{2}\langle \psi, e_i \cdot e_j \cdot \psi \rangle, \qquad \text{for } i < j. \qquad (3.1)$$

Here $F_A^+ = (1/2)F_{2A}^+$ where F_{2A}^+ is the self-dual part of the curvature of the connection d_{2A}. In the expression $e_i \cdot e_j \cdot \psi$, Clifford multiplication is understood.

We will sometimes write (d_A, ψ) instead of (d_{2A}, ψ) and think of d_A as a connection in the line bundle L, a bundle which strictly speaking does not exist unless M is spin. Often, we will simplify yet further, and write (A, ψ) instead of (d_A, ψ).

Equations (3.1) are known as the *Seiberg-Witten equations*. It follows from (2.7) that they can also be written in the form

$$D_A^+ \psi = 0, \qquad F_A^+ = \sigma(\psi). \qquad (3.2)$$

We will also need the "perturbed" Seiberg-Witten equations,

$$D_A^+ \psi = 0, \qquad F_A^+ = \sigma(\psi) + \phi, \qquad (3.3)$$

in which ϕ is a given self-dual two-form. Note that equations (3.2) and (3.3) fail to be linear only because of the presence of the term $\sigma(\psi)$, which is quadratic in ψ. This nonlinearity is much milder than that encountered in the Yang-Mills equations from nonabelian gauge theory.

Just like the Yang-Mills equations, the Seiberg-Witten equations are related to a variational principle. We define a nonnegative real-valued functional on the space

$$\mathcal{A} = \{(A, \psi) : \ A \text{ is a unitary connection on } L, \psi \in \Gamma(W_+ \otimes L)\},$$

by the formula

$$S(A, \psi) = \int_M \left[|D_A \psi|^2 + |F_A^+ - \sigma(\psi)|^2 \right] dV, \qquad (3.4)$$

where dV denotes the volume element on M. Clearly the absolute minimum of this functional is assumed at solutions to (3.2) when these solutions exist.

It follows from the fact that D_A is self-adjoint and from the Weitzenböck formula that

$$S(A, \psi) = \int_M \left[|D_A \psi|^2 + |F_A^+|^2 - 2\langle F_A^+, \sigma(\psi)\rangle + |\sigma(\psi)|^2 \right] dV$$

$$= \int_M \left[|D_A \psi|^2 + |F_A^+|^2 + \sum_{i<j} F_A^+(e_i, e_j)\langle \psi, ie_i e_j \psi\rangle + |\sigma(\psi)|^2 \right] dV$$

$$= \int_M \left[|\nabla^A \psi|^2 + \frac{s}{4}|\psi|^2 + |F_A^+|^2 + |\sigma(\psi)|^2 \right] dV.$$

In particular, if the scalar curvature s is strictly positive, all terms in the last expression are nonnegative and there are no nonzero solution to the Seiberg-Witten equations. Moreover, since

$$|\sigma(\psi)|^2 = \frac{1}{2}|\psi|^4,$$

if (A, ψ) is a solution to the Seiberg-Witten equations,

$$\int_M |F_A^+|^2 dV \leq \int_M \left[-\frac{s}{4}|\psi|^2 - \frac{1}{2}|\psi|^4 \right] dV \leq \int_M \frac{s^2}{32} dV. \qquad (3.5)$$

In the next few sections, we will study the properties of the space of solutions to the Seiberg-Witten equations. We will see that in the generic case, the moduli space of solutions to the perturbed Seiberg-Witten equations is a compact finite-dimensional manifold.

Remark: Several conventions are possible. For example, if (A, ψ) is a solution to (3.2), then $(A, c\psi)$ is a solution to the equations

$$D_A^+ \psi = 0, \qquad F_A^+ = c^2 \sigma(\psi),$$

and conversely, whenever c is a positive constant. Thus the theory of the latter equations is completely equivalent to that of the former, and which constant is used is merely a matter of taste.

3.2 The moduli space

We choose a "base connection" d_{A_0} in L so that our configuration space becomes

$$\mathcal{A} = \{(d_{A_0} - ia, \psi) : a \in \Omega^1(M), \psi \in \Gamma(W_+ \otimes L)\}$$

As in §1.7, the group of gauge transformations,

$$\mathcal{G} = \mathrm{Map}(M, S^1) = \{\text{ maps } g : M \to S^1\}$$

acts on $\mathcal{A}$ by

$$(g, (d_{A_0} - ia, \psi)) \mapsto (d_{A_0} - ia + gd(g^{-1}), g\psi).$$

In the case where M is simply connected, each element $g \in \mathcal{G}$ has a global logarithm u, so we can write

$$g = e^{iu}, \qquad u : M \to \mathbb{R},$$

and the action of $\mathcal{G}$ on $\mathcal{A}$ simplifies to

$$(g, (d_{A_0} - ia, \psi)) \mapsto (d_{A_0} - i(a + du)), e^{iu}\psi).$$

As we saw in §1.7, $\mathcal{G}$ possesses a subgroup of "based gauge transformations,"

$$\mathcal{G}_0 = \{g \in \mathcal{G} : g(p_0) = 1\},$$

p_0 being a chosen basepoint in M. The importance of the group $\mathcal{G}_0$ is that it acts freely on $\mathcal{A}$. Let $\widetilde{\mathcal{B}} = \mathcal{A}/\mathcal{G}_0$.

Proposition. *If M is simply connected, each element of $\widetilde{\mathcal{B}}$ has a unique representative of the form*

$$(d_{A_0} - ia, \psi), \qquad \text{where} \quad \delta a = 0.$$

Proof of existence: It suffices to find a function $u : M \to \mathbb{R}$ such that $\delta(a + du) = 0$. In other words, it suffices to find an element $u \in \Omega^0(M)$ such that

$$\Delta u = \delta(du) = -\delta a, \tag{3.6}$$

which is simply Poisson's equation. To solve it, we use Hodge theory: The theorem of Stokes shows that

$$\int_M 1 \wedge \star(\delta a) = \int_M d(\star a) = 0,$$

so δa lies in the orthogonal complement of the space $\mathcal{H}^0(M)$ of constant functions on M, with respect to the L^2 inner product ( ). Since this is also the orthogonal complement to the range of Δ by Hodge's Theorem, there does indeed exist a solution u to (3.6).

Proof of uniqueness: If

$$(d_{A_0} - ia_1, \psi_1), \qquad (d_{A_0} - ia_2, \psi_2)$$

are two elements of $\mathcal{A}$ which are equivalent under the action of $\mathcal{G}_0$ such that δa_1 and δa_2 are both zero, then since $da_1 = da_2$ and M is simply connected, there exists a map $u : M \to \mathbb{R}$ such that $a_1 - a_2 = du$. Hence

$$(a_1 - a_2, a_1 - a_2) = (du, a_1 - a_2) = (u, \delta(a_1 - a_2)) = 0 \quad \Rightarrow \quad a_1 - a_2 = 0.$$

Thus in the simply connected case, $\widetilde{\mathcal{B}}$ is in one-to-one correspondence with a linear subspace of $\mathcal{A}$:

$$\widetilde{\mathcal{B}} \cong \{(d_{A_0} - ia, \psi) : a \in \Omega^1(M), \psi \in \Gamma(W_+ \otimes L), \delta a = 0\}. \qquad (3.7)$$

However, we actually want to divide out by the full gauge group, and now we encounter an essential difference between the Seiberg-Witten theory and abelian gauge theory—instead of acting trivially, the group $U(1)$ of constant gauge transformations acts freely on $\widetilde{\mathcal{B}}$ except at the points $(d_{A_0} - ia, 0)$. The quotient space $\mathcal{B} = \mathcal{A}/\mathcal{G}$ by the full gauge group will have singularities at the "reducible" elements $(d_{A_0} - ia, 0)$. Sometimes, we can avoid dealing with the reducible elements by excising these singular points and setting

$$\mathcal{A}^* = \{(d_{A_0} - ia, \psi) \in \mathcal{A} : \psi \neq 0\}, \qquad \widetilde{\mathcal{B}}^* = \mathcal{A}^*/\mathcal{G}_0, \qquad \mathcal{B}^* = \mathcal{A}^*/\mathcal{G}.$$

We want to think of $\mathcal{B}^*$ as an infinite-dimensional manifold modeled on a Hilbert or Banach space, as described in [24]. To do this precisely, we need to complete our spaces of sections—just as in the old Yang-Mills theory—with respect to suitable Sobolev norms. We give only the basic definitions here and refer the reader to [16] or [14] for a more detailed treatment of "Sobolev completions."

If E is a smooth $O(m)$- or $U(m)$-bundle over a compact Riemannian manifold M with connection d_A, we can use the Levi-Civita connection on TM to define connections (also denoted by d_A) on the bundles $\otimes^k T^*M \otimes E = \mathrm{Hom}(\otimes^k TM, E)$. If $\sigma \in \Gamma(E)$, we can define

$$d_A^k \sigma = (d_A \circ \cdots \circ d_A)\sigma \in \Gamma(\mathrm{Hom}(\otimes^k TM, E)).$$

For $p > 1$, let

$$\|\sigma\|_{p,k} = \left[\int_M [|\sigma|^p + |d_A \sigma|^p + \cdots + |d_A^k \sigma|^p] dV \right]^{(1/p)}.$$

It can be checked that this is a norm on $\Gamma(E)$, and we let $L_k^p(E)$ denote the completion of $\Gamma(E)$ with respect to this norm. The choice of Riemannian metric, the fiber metric on E or the connection d_A replaces the norms on $\Gamma(E)$ by equivalent norms, and hence does not affect the resulting completions.

For all choices of p, $L_k^p(E)$ is a Banach space, and it is a Hilbert space if $p = 2$. The spaces $L_k^p(E)$ are called *Sobolev spaces*.

Here are some key results from functional analysis (which are described in more detail in the appendix to Donaldson and Kronheimer [14]): In the case where the base manifold has dimension four, the Sobolev Embedding Theorem states that if $k - (4/p) > l$ there is a continuous embedding

$$L_k^p(E) \to C^l(E),$$

the space of C^l sections of E. Rellich's Theorem states that the inclusion

$$L_{k+1}^p(E) \to L_k^p(E)$$

is compact for all p and k, meaning that a sequence σ_i which is bounded in L_{k+1}^p possesses a subsequence which converges in L_k^p. When $k - (4/p) > 0$, the multiplication theorems state that there are continuous multiplications

$$L_k^p(E) \times L_k^p(F) \to L_k^p(E \otimes F).$$

Thus for p and k in this range and for E the trivial line bundle, $L_k^p(E)$ is a Banach algebra.

With these preparations out of the way, we can set

$$\widetilde{\mathcal{B}}_k^p = \{(d_{A_0} - ia, \psi) : a \in L_k^p(T^*M), \psi \in L_k^p(W_+ \otimes L), \delta a = 0\}.$$

Similarly, we can define $\mathcal{A}_k^p$,

$$\mathcal{G}_{k+1}^p = L_k^p(\mathrm{End}(E)) \cap C^0(M, S^1),$$

when $k + 1 - (4/p) > 0$, and so forth. It can also be shown that $\mathcal{G}_{k+1}^p$ is an infinite-dimensional Lie group which acts smoothly on $\mathcal{A}_k^p$, although these last facts will not be needed in our subsequent arguments.

Henceforth, when we write spaces such as $\mathcal{A}$, $\widetilde{\mathcal{B}}^*$ or $\mathcal{G}$, appropriate Sobolev completions will be understood. We will only include the indices p and k when explicit values are needed for clarity.

After Sobolev completion, $\widetilde{\mathcal{B}}^*$ can be regarded as an S^1 bundle over the infinite-dimensional manifold $\mathcal{B}^*$. Just as $\pi_k(S^n) = 0$ for $k < n$, the k-th homotopy group of an infinite-dimensional sphere must vanish for all k. Thus $\mathcal{A}^*$, which is homotopy equivalent to an infinite-dimensional linear space minus a point, must have the homotopy groups of a point. If M is simply connected, $\mathcal{G}_0$ is also contractible, and hence $\widetilde{\mathcal{B}}^*$ has the homotopy groups of a point. It follows from the exact homotopy sequence of the fiber bundle

$$S^1 \to \widetilde{\mathcal{B}}^* \to \mathcal{B}^* \tag{3.8}$$

that $\mathcal{B}^*$ is a $K(\mathbb{Z}, 2)$, just like $P^\infty \mathbb{C}$. In fact, all $K(\mathbb{Z}, 2)$'s are homotopy equivalent, so $\mathcal{B}^*$ is homotopy equivalent to $P^\infty \mathbb{C}$.

We can define a complex line bundle E over $\mathcal{B}^*$ which pulls back to the universal bundle under the homotopy equivalence $P^\infty \mathbb{C} \to \mathcal{B}^*$. Then $\widetilde{\mathcal{B}}^*$ can be identified with the bundle of unit-length vectors in E.

Definition. The *monopole moduli space* is

$$\mathcal{M} = \{[A, \psi] \in \mathcal{B} : (A, \psi) \text{ satisfies (3.2)}\},$$

or more generally, if ϕ is a given self dual two-form,

$$\mathcal{M}_\phi = \{[A, \psi] \in \mathcal{B} : (A, \psi) \text{ satisfies (3.3) }\}.$$

Using (3.7). we see that the latter space is the quotient of

$$\widetilde{\mathcal{M}}_\phi = \{(d_{A_0} - ia, \psi) \in \Omega^1(M) \times \Gamma(W_+ \otimes L) :$$

$$\delta a = 0 \text{ and } (d_{A_0} - ia, \psi) \text{ satisfies (3.3) }\}$$

by the action of S^1. We will see that $\widetilde{\mathcal{M}}_\phi$ is a finite-dimensional submanifold of the linear space $\widetilde{\mathcal{B}}$ except possible for singularities at points where $\psi = 0$.

3.3 Compactness of the moduli space

The first remarkable property of the moduli spaces $\mathcal{M}$ or $\mathcal{M}_\phi$ is that they are compact, as we now show following the treatment in Kronheimer and Mrowka [23].

Lemma [23]. *If (A, ψ) is a solution to the Seiberg-Witten equations with ψ not identically zero, and the maximum value of $|\psi|$ is assumed at a point $p \in M$, then*

$$|\psi|^2(p) \leq -\frac{1}{4} s(p),$$

where s is the scalar curvature.

Proof: The restriction of the Hodge Laplacian to zero-forms or functions can be expressed as

$$\Delta = -\frac{1}{\sqrt{g}} \sum_{i,j=1}^{4} \frac{\partial}{\partial x_i} \left(g^{ij} \frac{\partial}{\partial x_j} \right),$$

where g_{ij} are the components of the meric with respect to coordinates $(x_1, \ldots, x_4)$ and $(g^{ij}) = (g_{ij})^{-1}$. Note that $\Delta(f) \geq 0$ at a local maximum of f, and hence since p is a maximum for $|\psi|^2$,

$$\frac{1}{2}\Delta(|\psi|^2)(p) \geq 0.$$

Since d_A is a metric connection,

$$-\langle d_A\psi, d_A\psi \rangle(p) + \mathrm{Re}\langle \psi, \Delta^A\psi \rangle(p) \geq 0 \quad \Rightarrow \quad \mathrm{Re}\langle \psi, \Delta^A\psi \rangle(p) \geq 0,$$

where Δ^A is the vector bundle Laplacian defined by (2.13). It follows from the Weitzenböck formula that

$$0 = D_A^2\psi = \Delta^A\psi + \frac{s}{4}\psi - \sum_{i<j} F_A(e_i, e_j)(ie_i \cdot e_j \cdot \psi).$$

Take the inner product with ψ and apply the second of the Seiberg-Witten equations to obtain

$$-\frac{s}{4}|\psi|^2 - \frac{1}{2}\sum_{i<j}|F_A^+(e_i, e_j)|^2 = \langle \psi, \Delta^A\psi \rangle(p) \geq 0.$$

Thus

$$-\frac{s(p)}{2}|\psi(p)|^2 \geq |F_A^+(p)|^2 = 2|\psi(p)|^4.$$

Divide by $|\psi(p)|^2$ to obtain the desired result.

The Lemma shows that the range of ψ is contained in a compact subset of $W_+ \otimes L$. To show that the moduli space itself is compact, however, we need to use the Sobolev completions described in the preceding section, as well as the following key inequality from PDE theory: If D is a Dirac operator or a Dirac operator with coefficients (such as $d + \delta$), then

$$\|\sigma\|_{p,k+1} \leq c[\|D\sigma\|_{p,k} + \|\sigma\|_{p,k}], \tag{3.9}$$

where c is a constant. The $\|\sigma\|_{p,k}$ term can be omitted in the case where D has trivial kernel. In the case where $p = 2$ inequality (3.9) is often used to prove Hodge's Theorem. The general case is discussed in the Appendix to [14] and Appendix B to [27].

Compactness Theorem. *If M is simply connected, then for every choice of self-dual two-form ϕ, the moduli space $\widetilde{\mathcal{M}}_\phi$ of solutions to the perturbed Seiberg-Witten equations is compact.*

Remark: Simple connectedness of M is not actually necessary, but makes the proof especially simple.

Proof: We choose a base connection A_0, so that any unitary connection on L is of the form $d_{A_0} - ia$, for some $a \in \Omega^1(M)$. We need to show that any sequence $[d_{A_0} - ia_i, \psi_i]$ of solutions to the perturbed Seiberg-Witten equations possesses a convergent subsequence.

Imposing the gauge condition $\delta a = 0$ puts the perturbed Seiberg-Witten equations into the form

$$\begin{cases} D_{A_0}\psi - ia \cdot \psi = 0, \\ (da)^+ + F_{A_0}^+ = \sigma(\psi) + \phi, \\ \delta a = 0, \end{cases}$$

or equivalently,

$$\begin{cases} D_{A_0}\psi = ia \cdot \psi, \\ (da)^+ = \sigma(\psi) + \phi - F_{A_0}^+, \\ \delta a = 0. \end{cases}$$

On the left-hand side of this system appear the two Dirac operators D_{A_0} and $(\delta \oplus d^+)$, the second of these being the Dirac operator with coefficients in W_+ described in §2.7. The Lemma implies that ψ_i is bounded in C^0 and hence in every L^p. Since M is simply connected, the operator

$$\delta \oplus d^+ : \Omega^1(M) \to \Omega^0(M) \oplus \Omega_+^2(M)$$

has trivial kernel. It therefore follows from (3.9) applied to the last two equations that a_i is bounded in L_1^p for all p. In particular, taking $p > 4$, we see that it is bounded in C^0 by the Sobolev Embedding Theorem, and hence $a_i\psi_i$ is bounded in L^p for all p. Then it follows from (3.9) applied to the first equation that ψ_i is bounded in L_1^p for all p. If $p > 4$, L_1^p is in Banach algebra range, so

$$a_i, \psi_i \text{ bounded in } L_1^p \;\Rightarrow\; a_i\psi_i, \sigma(\psi_i) \text{ bounded in } L_1^p$$

$$\Rightarrow\; a_i, \psi_i \text{ bounded in } L_2^p.$$

Similarly,

$$a_i, \psi_i \text{ bounded in } L_k^p \;\Rightarrow\; a_i\psi_i, \sigma(\psi_i) \text{ bounded in } L_k^p$$

$$\Rightarrow \ a_i, \psi_i \text{ bounded in } L^p_{k+1}.$$

Thus we can "bootstrap" to conclude that a_i, ψ_i are bounded in L^p_k for all k. Then Rellich's Theorem produces a subsequence which converges in L^p_k for all k, and the sequence must converge in C^l for all l by the Sobolev Embedding Theorem. This proves compactness.

3.4   Transversality

The second remarkable property of the moduli space $\mathcal{M}_\phi$ is that for generic choice of ϕ it is a smooth manifold, except possibly at reducible elements. To prove this, we need to use Smale's infinite-dimensional version of Sard's theorem [36].

A familiar method for constructing smooth submanifolds of a finite-dimensional smooth manifold M^m goes like this: Suppose that $F : M^m \rightarrow N^n$ be a smooth map which has q as a regular value. Then $F^{-1}(q)$ is a smooth embedded submanifold of M^m by the implicit function theorem, and its dimension is $m - n$. This is complemented by Sard's theorem, which states that almost all $q \in N^n$ are regular values.

In order to apply this idea to nonlinear PDE's, we need to generalize to maps between manifolds modeled on infinite-dimensional Banach spaces. To do this we need Smale's notion of Fredholm map. Suppose that E_1 and E_2 are reflexive Banach spaces. A continuous linear map $T : E_1 \rightarrow E_2$ is said to be *Fredholm* if

1. the kernel of T is finite-dimensional,

2. the range of T is closed, and

3. the range of T has finite codimension in E_2.

The *index* of a Fredholm map T is defined to be

$$\dim(\mathrm{Ker}(T)) - \mathrm{codim}(T(E_1)) = \dim(\mathrm{Ker}(T)) - \dim(\mathrm{Ker}(T^*)),$$

where $T^* : E_2^* \rightarrow E_1^*$ is the adjoint of T defined by $T^*(\phi)(\psi) = \phi(T(\psi))$. A key feature of Fredholm operators is that their index is invariant under perturbation; if a family depends continuously upon a parameter which ranges through a connected topological space, the index of the family must be constant.

For example, if M is a compact four-manifold and

$$D_A^+ : \Gamma(W_+ \otimes L) \rightarrow \Gamma(W_- \otimes L)$$

is a Dirac operator with coefficients in a line bundle L, D_A^+ extends to a continuous linear map

$$D_A^+ : L_{k+1}^p(W_+ \otimes L) \to L_k^p(W_- \otimes L),$$

for every choice of p and $k \geq 0$. This extension is a Fredholm map with adjoint D_A^- ([25], Theorem 5.2, page 193) and its Fredholm index is twice the complex index defined before, which is calculated by the Atiyah-Singer Index Theorem.

Suppose now that $F : M_1 \to M_2$ is a smooth *nonlinear* map from a Banach manifold M_1 to a Banach manifold M_2. We say that F is *Fredholm* of index k if its linearization at p,

$$dF(p) : T_p M_1 \to T_{F(p)} M_2,$$

is Fredholm map of index k for every $p \in M_1$. Note that a smooth map $F : M_1 \to M_2$ between *finite-dimensional* manifolds is automatically a Fredholm map of index $\dim M_1 - \dim M_2$.

A point $q \in M_2$ is said to be a *regular value* of F if $dF(p)$ is surjective for every $p \in F^{-1}(q)$; otherwise it is called a *critical value*. It follows from the implicit function theorem on Banach spaces (see [24]) that if $q \in M_2$ is a regular value, then $F^{-1}(q)$ is a smooth submanifold of M_1, just as in the finite-dimensional case. Moreover, $F^{-1}(q)$ is finite-dimensional, its dimension being the Fredholm index of F.

Finally, a subset of the Banach manifold M_2 is called *residual* if it is a countable intersection of open dense sets. Recall that the Baire Category Theorem states that a residual subset of a complete metric space is dense. It is customary to refer to an element of a residual subset of M_2 as a *generic element*.

With these preparations out of the way, we can now state Smale's extension of Sard's theorem [36]:

Sard-Smale Theorem. *If $F : M_1 \to M_2$ is a C^k Fredholm map between separable Banach manifolds and $k > \max(0, \text{index of } F)$, then the set of regular values of F is residual in M_2.*

In other words, a generic element of M_2 is a regular value. In our applications, the separability assumption will automatically hold, since all of the manifolds modeled on L_k^p spaces that we have constructed are separable. The idea behind the proof of the Sard-Smale Theorem is to reduce to the finite-dimensional version of Sard's theorem; we refer the reader to [36] for further details.

We will need one further theorem from the theory of elliptic PDE's, the Unique Continuation Theorem found in [2] and §8 of [7]. This theorem asserts that if ψ is a solution to the equation $D_A^+\psi = 0$ on a connected manifold, then ψ cannot vanish on an open set without vanishing identically. This is analogous to a familiar property of holomorphic functions. (Indeed, the Dirac operator can be thought of as a generalization of the Cauchy-Riemann operator from two to four dimensions.)

As a first step towards the main theorems of this section, we present the following simple application of the Sard-Smale Theorem:

Lemma. *Suppose that the index of the Dirac operator D_A^+ is nonnegative, as A ranges through the space of unitary connections on L. Then for a generic choice of A,*

$$D_A^+ : \Gamma(W_+ \otimes L) \to \Gamma(W_- \otimes L)$$

is surjective.

Proof: Define

$$F : \mathcal{A} \to \Gamma(W_- \otimes L) \qquad \text{by} \qquad F(A, \psi) = D_A^+\psi.$$

The differential of F at (A, ψ) is

$$dF(A, \psi)(a, \psi') = D_A^+\psi' - ia \cdot \psi,$$

the dot denoting Clifford multiplication between the one-form a and the spinor field ψ. We claim that if (A, ψ) is a solution to the Seiberg-Witten equations with $\psi \neq 0$, then $dF(A, \psi)$ is surjective.

Note first that at points where $\psi \neq 0$, the linear map $a \mapsto a \cdot \psi$ is injective (because $a \cdot a \cdot \psi = -|a|^2\psi$), and hence it is an isomorphism (since T^*M and $W_+ \otimes L$ have the same dimension). If ψ is not identically zero, it must be nonzero on an open set U, and the map $a \mapsto a \cdot \psi$ must be an isomorphism for sections supported in this set.

We choose $p \geq 2$ and $k \geq 0$ so that $L_k^p(W_- \otimes L)$ is a dense linear subspace of $L^2(W_- \otimes L)$. Thus to show that $dF(A, \psi)$ is surjective, it suffices to check that if $\sigma \in L^2(W_- \otimes L)$ is perpendicular to the image of $dF(A, \psi)$, it must be zero. But in view of the previous paragraph, this condition implies that $(\tau, \sigma) = 0$ for all elements $\tau \in L^2(W_- \otimes L)$ which are supported in U, where $(\cdot, \cdot)$ denotes the L^2 inner product. Since $D_A^-\sigma = 0$ by self-adjointness of the Dirac operator, it follows from the Unique Continuation Theorem that σ is identically zero. Thus $dF(A, \psi)$ is indeed surjective and our claim is established.

The implicit function theorem now implies that

$$\mathcal{N} = \{(A, \psi) \in \widetilde{\mathcal{B}}_k^p : D_A^+\psi = 0, \psi \neq 0\}$$

is a submanifold of $\widetilde{\mathcal{B}}_k^p$ whose tangent space at the point (A, ψ) is

$$T_{(A,\psi)}\mathcal{N} = \{(a, \psi') : D_A^+\psi' - ia \cdot \psi = 0\}.$$

We claim that the map

$$\pi : \mathcal{N} \to \{ \text{ unitary connections on } L \}, \qquad \pi(A, \psi) = A,$$

is Fredholm. Indeed, the kernel of $d\pi$ is the set of (a, ψ') in the tangent space which satisfy the condition $a = 0$, which is just the kernel of D_A^+. On the other hand, the image of $d\pi$ includes all one-forms a such that $a \cdot \psi$ is in the image of D_A^+, a closed subspace of finite codimension, the codimension being $\leq \dim(\text{Ker}(D_A^-))$. Thus π is Fredholm and its Fredholm index is at least as large as the index of D_A^+.

Choose A to be a regular value of π. Then $\pi^{-1}(A)$ is a submanifold whose dimension is at least as large as the index of D_A^+, while $\pi^{-1}(A) \cup \{(A, 0)\}$ is the the set of pairs (A, ψ) such that ψ lies in the linear space of solutions to $D_A^+\psi = 0$. The definition of π shows that for a generic choice of connection A, D_A^+ is surjective, and the lemma is proven.

Transversality Theorem 1. *Let M be a compact simply connected smooth four-manifold with a spin^c-structure. In the special case in which $b_+(M) = 0$, we make the additional assumption that the index of the Dirac operator D_A^+ associated to the spin^c-structure is ≥ 0. Then for a generic choice of self-dual two-form ϕ, $\widetilde{\mathcal{M}}_\phi$ is an oriented smooth manifold, whose dimension is given by the formula*

$$\dim(\widetilde{\mathcal{M}}_\phi) = 2(\text{index of } D_A^+) - b_+. \qquad (3.10)$$

Proof: Except for orientation, the proof is similar to the proof of the previous lemma. Define

$$F : \widetilde{\mathcal{B}} \times \Omega_+^2(M) \to \Gamma(W_- \otimes L) \times \Omega_+^2(M)$$

by

$$F(A, \psi, \phi) = (D_A^+\psi, F_A^+ - \sigma(\psi) - \phi).$$

The differential of F at (A, ψ, ϕ) is

$$dF(A, \psi, \phi)(a, \psi', \phi') = (D_A^+\psi' - ia \cdot \psi, (da)^+ - 2\sigma(\psi, \psi') - \phi'),$$

where

$$\sigma(\psi, \psi') = 2i \text{ Trace-free Hermitian part of } \begin{pmatrix} \psi_1 \\ \psi_2 \end{pmatrix} (\bar{\psi}_1' \quad \bar{\psi}_2').$$

We claim that if (A, ψ, ϕ) is a solution to the equation $F = 0$ with $\psi \neq 0$, then $dF(A, \psi, \phi)$ is surjective. Indeed, by letting ϕ' vary, we see that the image of $dF(A, \psi, \phi)$ is foliated by translates of the linear space $\{0\} \oplus \Omega_+^2(M)$. Thus it suffices to show that the first component is surjective when we set $\phi' = 0$. Suppose that $(\sigma, 0)$ is an L^2 section of $(W_- \otimes L) \oplus \Lambda_+^2$ which is perpendicular to

$$dF(A, \psi, \phi)(T_{[A, \psi]}\widetilde{B} \times \{0\}).$$

By letting a vary, we can argue as in the proof of the lemma that σ is orthogonal to all sections of $W_- \otimes L$ which are supported in the nonempty open set $U = \{p : \psi(p) \neq 0\}$, so σ must vanish on U. By letting ψ' vary, we see that $D_A^- \sigma = 0$, so the Unique Continuation Theorem once again implies that σ is identically zero. Thus $dF(A, \psi, \phi)$ is surjective, except possibly at points where $\psi = 0$.

We now set

$$\mathcal{U} = \{\phi \in \Omega_+^2 : \text{ if } \phi = F_A^+ \text{ for a connection } A, \text{ then } D_A^+ \text{ is surjective }\},$$

so that $\phi \in \mathcal{U}$ implies that $dF(A, \psi, \phi)$ is surjective, even if $\psi = 0$. Note that if $b_+ > 0$, $\mathcal{U}$ contains the complement of the affine subspace Π of codimension b_+ described in the Proposition from §1.9 and is therefore open and dense. On the other hand, if $b_+ = 0$, the hypothesis that the index of D_A^+ is nonnegative ensures that $\mathcal{U} = \Omega_+^2$ by the preceding lemma. In either case, $\mathcal{U}$ is open and dense in Ω_+^2. Moreover, by the implicit function theorem,

$$\mathcal{N} = \{(A, \psi, \phi) \in \widetilde{B} \times \mathcal{U} : F(A, \psi, \phi) = 0\}$$

is a submanifold of $\widetilde{B} \times \mathcal{U}$ whose tangent space at the point (A, ψ, ϕ) is

$$T_{(A, \psi, \phi)}\mathcal{N} = \{(a, \psi', \phi') : L(a, \psi') = (0, 0, \phi')\},$$

where

$$L(a, \psi') = (D_A^+ \psi' - ia \cdot \psi, \delta a, (da)^+ - 2\sigma(\psi, \psi')).$$

In this last formula, L is an elliptic operator,

$$L : \Gamma(W_+ \otimes L) \oplus \Omega^1(M) \to \Gamma(W_- \otimes L) \oplus \widetilde{\Omega}^0(M) \oplus \Omega_+^2(M),$$

the symbol $\widetilde{\Omega}^0(M)$ denoting the space of smooth functions on M which integrate to zero.

We claim that the projection

$$\pi : \mathcal{N} \to \Omega_+^2(M), \qquad \text{defined by} \qquad \pi(A, \psi, \phi) = \phi,$$

is a Fredholm map. Indeed, the kernel of $d\pi$ is immediately seen to be the kernel of L, while

$$(\text{Image of } d\pi) = \{\phi' \in \Omega_+^2 : (0,0,\phi') = L(a,\psi') \text{ for some } a, \psi' \}$$

$$= (\text{Image of } L) \cap (0 \oplus 0 \oplus \Omega_+^2),$$

so the image of $d\pi$ is closed and of finite codimension.

To show that the cokernel of $d\pi$ has the same dimension as the cokernel of L, it suffices to show that

$$(\Gamma(W_- \otimes L) \oplus \widetilde{\Omega}^0(M) \oplus \{0\}) \cap (\text{Image of } L)^\perp = \{0\}. \tag{3.11}$$

But if

$$(\sigma, u, 0) \in \Gamma(W_- \otimes L) \oplus \widetilde{\Omega}^0(M) \oplus \Omega_+^2(M)$$

is perpendicular to the image of L, then $D_A^- \sigma = 0$, and $(ia \cdot \psi, \sigma) = (\delta a, u)$, for all $a \in \Omega^1(M)$. In particular, $(ib \cdot \psi, \sigma) = 0$ for all b such that $\delta b = 0$. On the other hand, we can define a one-form b on M by

$$\langle b, a \rangle = \langle ia \cdot \psi, \sigma \rangle.$$

and since $D_A^+ \psi = 0$ and $D_A^- \sigma = 0$, it follows from (2.15) that $\delta b = 0$. Thus

$$(b,b) = (ib \cdot \psi, \sigma) = 0 \qquad \Rightarrow \qquad b = 0.$$

If $b_+ > 0$, we can choose ϕ to lie in the complement of the affine space Π, so that

$$(A, \psi) \in \widetilde{\mathcal{M}}_\phi \qquad \Rightarrow \qquad \psi \neq 0.$$

It then follows from the fact that $\langle ia \cdot \psi, \sigma \rangle = 0$ for all $a \in \Omega^1(M)$ that $\sigma = 0$ on a nonempty open subset of M. By the Unique Continuation Theorem once again, $\sigma \equiv 0$ and hence $(\delta a, u) = 0$ for all $a \in \Omega^1$. Thus $u = 0$ and (3.11) holds. If $b_+ = 0$ and $\psi = 0$, the operator L simplifies to $D_A^+ \oplus (\delta \oplus d^+)$; in this case the first component is surjective by the Lemma, while the second component is surjective by Hodge theory. Thus L itself is surjective and (3.11) holds once again.

Choose ϕ to be a regular value of π. Then $\pi^{-1}(\phi)$ is a submanifold of $\mathcal{N}$ whose dimension is the index of L, which is the same as the index of

$$L_0 = D_A^+ \oplus \delta \oplus d^+.$$

The index we want is a real index, so the index of D_A^+ given by the Atiyah-Singer index must be doubled in the present calculation. We find that

$$\text{real index of } L = 2(\text{complex index of } D_A^+) - b_+$$

$$= -\frac{\tau(M)}{4} + c_1^2(L)[M] - b_+.$$

Thus the moduli space $\widetilde{\mathcal{M}}_\phi = \pi^{-1}(\phi)$ is indeed a smooth submanifold of the required dimension.

Orientation of the moduli space: To finish the proof of Transversality Theorem 1, we need only show that the moduli space $\widetilde{\mathcal{M}}_\phi$ is oriented. This requires some results from the theory of Fredholm operators (presented, for example, in the Appendix to [3] or in Chapter III, §7 of [25]). These results are incorporated in the notion of determinant line bundle for a family of Fredholm operators which depend upon a parameter which ranges through a smooth manifold.

To make this precise, we suppose that $p \mapsto L(p)$ is a family of Fredholm operators on a real Hilbert space H, say $L(p) : H \to H$, depending continuously on $p \in M$, where M is a finite-dimensional smooth manifold. Then by Proposition A.5 on page 156 of [3], there is a closed linear subspace $V \subset H$ of finite codimension such that

1. V does not intersect the kernel of $L(p)$ for any $p \in M$.

2. $(L(p)(V))$ has constant codimension and $p \mapsto H/(L(p)(V))$ is a finite-dimensional vector bundle E over M.

3. The linear maps $L(p)$ induce a vector bundle map $\tilde{L} : (H/V) \times M \to E$.

We can then define the *determinant line bundle* of L by

$$\det(L) = \Lambda^{\max}((H/V) \times M) \otimes [\Lambda^{\max}(E)]^*,$$

where max denotes the rank of the vector bundle in question.

We claim that the determinant line bundle is independent of the choice of V. Indeed, if V_1 is a second choice of linear subspace satisfying the above conditions, we can assume without loss of generality that $V_1 \subset V$, and then

$$\frac{H}{V_1} \cong \frac{H}{V} \oplus \frac{V}{V_1}, \qquad E \cong E_1 \oplus \frac{V}{V_1},$$

where E_1 is the line bundle whose fiber at p is $H/(L(p)(V))$. Since

$$\Lambda^{\max}(W_1 \oplus W_2) \cong \Lambda^{\max}(W_1) \otimes \Lambda^{\max}(W_1),$$

$$\Lambda^{\max}(W) \otimes \Lambda^{\max}(W^*) \cong \Theta,$$

Θ being the trivial line bundle, we see that

$$\Lambda^{\max}((H/V_1) \times M) \otimes [\Lambda^{\max}(E_1)]^* \cong \Lambda^{\max}((H/V) \times M) \otimes [\Lambda^{\max}(E)]^*.$$

Now we can use the exact sequence

$$0 \to \mathrm{Ker}(L(p)) \to \frac{H}{V} \to \frac{H}{L(p)(V)} \to \mathrm{Coker}(L(p)) \to 0$$

to show that there is a natural isomorphism

$$\det(L)(p) \cong \Lambda^{\max}(\mathrm{Ker}(L(p))) \otimes [\Lambda^{\max}(\mathrm{Coker}(L(p)))]^*.$$

The key point of this construction is that the determinant line bundle of a family of elliptic operators L is well-defined even though the dimensions of the kernels and cokernels vary from point to point.

It is not difficult to extend the above discussion to families of Fredholm maps, $L(p) : H_1 \to H_2$, between two different Hilbert spaces. In our case, we have a family of *surjective* linear operators $(A, \psi) \in \widetilde{\mathcal{M}}_\phi \mapsto L_{A,\psi}$ which are Fredholm on the L_k^2 completions, with

$$\mathrm{Ker}(L_{A,\psi}) = T_{A,\psi}(\widetilde{\mathcal{M}}_\phi).$$

Hence if d is the dimension of the moduli space, the determinant line bundle of L is the top exterior power of the tangent bundle to $\widetilde{\mathcal{M}}_\phi$,

$$\det(L) = \Lambda^d \mathrm{Ker}(L) = \Lambda^d(\widetilde{\mathcal{M}}_\phi).$$

Thus a nowhere zero section of $\det(L)$ will give an orientation of $\widetilde{\mathcal{M}}_\phi$.

If we define a family of elliptic operators L_t, for $t \in [0,1]$, by

$$L_t(a, \psi') = (D_A^+ \psi' - ita \cdot \psi, \delta a, (da)^+ - 2t\sigma(\psi, \psi')),$$

the determinant line bundle $\det(L_t)$ is defined for every t and depends continuously on t. Thus the bundles $\det(L_t)$ are all isomorphic, and it suffices to construct a nowhere zero section of $\det(L_0)$, where $L_0 = D_A^+ \oplus \delta \oplus d^+$. But

$$\det(L_0) = \det(D_A^+) \otimes \det(\delta \oplus d^+).$$

The first factor has a nowhere zero section which comes from the orientations of the kernel and cokernel of D_A^+ defined by complex multiplication, whereas the second factor inherits a nowhere zero section from an orientation of $\mathcal{H}_+^2(M)$. Thus $\det(L_0)$ is trivialized, which in turn trivializes the top exterior power of the tangent space to the moduli space, thereby orienting the moduli space and finishing the proof of the theorem.

Transversality Theorem 2. *Let M be a compact simply connected smooth four-manifold with a spin^c-structure. If $b_+(M) > 0$, then for generic* choice of self-dual two-form ϕ, $\mathcal{M}_\phi$ is an oriented smooth manifold, whose dimension is given by the formula

$$\dim(\mathcal{M}_\phi) = 2(\text{index of } D_A^+) - b_+ - 1,$$

or equivalently

$$\dim(\mathcal{M}_\phi) = \langle c_1(L)^2, [M] \rangle - \frac{1}{4}(2\chi(M) + 3\tau(M)). \qquad (3.12)$$

Proof: Note that a solution to the perturbed Seiberg-Witten equations with $\psi = 0$ will occur only if $c_1(L)$ contains a connection with $F_A^+ = \phi$, and this will occur for ϕ lying in a subspace Π of $\Omega_+^2(M)$ of codimension $b_+ \geq$ one, by the Proposition of §1.9.

This theorem therefore follows from the preceding one by dividing out by the free S^1-action, which decreases the dimension by one. The second formula for the index follows from the Atiyah-Singer Index Theorem applied to D_A^+.

Remark 1: Note that for the moduli space to have nonnegative formal dimension,

$$c_1(L)^2[M] \geq \frac{\tau(M)}{4} - b_+ - 1,$$

and hence for any unitary connection on L,

$$\frac{1}{4\pi^2} \int_M [|F_A^+|^2 - |F_A^-|^2]dV \geq \frac{\tau(M)}{4} - b_+ - 1,$$

or

$$\int_M |F_A^-|^2 dV \leq \int_M |F_A^+|^2 dV - (4\pi^2)\left(\frac{\tau(M)}{4} - b_+ - 1\right). \qquad (3.13)$$

This, together with an extension of inequality (3.5) to the case of nonzero ϕ, shows that $\int_M |F_A|^2 dV$ is bounded, and hence for generic ϕ, the moduli space given by the Transversality Theorem 2 will be empty except for finitely many line bundles L.

Remark 2: As in the case of the Compactness Theorem, the Transversality Theorem 2 can be proven to hold without the assumption that M is simply connected, the dimension of the moduli space being given by (3.12).

3.5 The intersection form

We now describe some of the classical topological invariants of simply connected four-dimensional manifolds.

Let M be a compact simply connected four-manifold. The *intersection form* on integer cohomology is a symmetric bilinear map

$$Q : H^2(M; \mathbb{Z}) \times H^2(M; \mathbb{Z}) \to \mathbb{Z},$$

defined to be the composition of the cup product

$$H^2(M; \mathbb{Z}) \times H^2(M; \mathbb{Z}) \to H^4(M; \mathbb{Z})$$

with the standard isomorphism of $H^4(M; \mathbb{Z})$ to $\mathbb{Z}$ which pairs an element of $H^4(M; \mathbb{Z})$ with the orientation class of M.

The intersection form can also be defined on de Rham cohomology by integration of differential forms. Indeed, we can let

$$\widetilde{Q} : H^2(M; \mathbb{R}) \times H^2(M; \mathbb{R}) \to \mathbb{R} \qquad \text{by} \qquad \widetilde{Q}([\alpha], [\beta]) = \int_M \alpha \wedge \beta.$$

In the case where M is simply connected, it follows from the universal coefficient theorem that

$$H^2(M; \mathbb{Z}) \cong \mathrm{Hom}(H_2(M; \mathbb{Z}), \mathbb{Z}),$$

a free abelian group. This group lies inside the real cohomology as a lattice,

$$H^2(M; \mathbb{Z}) = \{[\alpha] \in H^2(M; \mathbb{R}) : \text{the integral}$$

of α over any compact surface in M is an integer}.

The restriction of $\widetilde{Q}$ to $H^2(M; \mathbb{Z})$ is integer-valued, and this restriction is the intersection form defined before.

Finally, we can also regard the intersection form as defined on integer homology $H_2(M; \mathbb{Z})$. To see this, we need to utilize an important fact from the topology of four-manifolds: any element of $H_2(M; \mathbb{Z})$ can be represented by a compact oriented surface Σ embedded in M. One way of proving this is to note that since M is assumed to be simply connected, it follows from the Hurewicz isomorphism theorem that $H_2(M; \mathbb{Z}) \cong \pi_2(M)$ and hence any element of $H_2(M; \mathbb{Z})$ can be represented by a smooth map $f : S^2 \to M$. We can arrange, after a possible perturbation, that this map be an immersion with transverse self-intersections, and the self-intersections can be removed by surgeries, each of which replaces two intersecting disks by an annulus. The result is that a given element of $H_2(M; \mathbb{Z})$ can be represented by an embedded surface which may have several components, each of which may have arbitrary genus.

Using this fact, we can define a map

$$\mu : H_2(M; \mathbb{Z}) \to H^2(M; \mathbb{Z})$$

as follows: An element of $H_2(M; \mathbb{Z})$ can be represented by a compact oriented surface Σ. This surface Σ lies within a small tubular neighborhood V, the image under the exponential map of an ϵ-neighborhood of the zero section in the normal bundle of Σ in M. The Thom form on the total space of the normal bundle, chosen as described in §1.6 to have compact support within the given ϵ-neighborhood, can be pushed forward under the exponential map to a closed two-form Φ_Σ which has compact support contained within V. Extending Φ_Σ by zero to all of M gives a closed two-form defined on M, which we still denote by Φ_Σ. If Σ' is an oriented embedded surface in M which has transverse intersection with Σ, the integral of Φ_Σ over Σ' will be an integer, since the integral of the Thom form over any fiber is an integer. In fact this integer is the intersection number of Σ with Σ' as described at the end of §1.6. Thus the de Rham cohomology class $[\Phi_\Sigma]$ of Φ_Σ lies in the image of $H^2(M; \mathbb{Z})$ under the coefficient homomorphism. We set

$$\mu([\Sigma]) = [\Phi_\Sigma] \in H^2(M; \mathbb{Z}).$$

By the Classification Theorem for Complex Line Bundles, an element of $H^2(M; \mathbb{Z})$ is the first Chern class of a complex line bundle L over M. Choose a smooth section $\sigma : M \to L$ which has transverse intersection with the zero section, and let

$$Z_\sigma = \sigma^{-1}(\text{zero section}).$$

If Φ_{Z_σ} is the differential form constructed in the previous paragraph, the integral of Φ_{Z_σ} over any compact oriented surface Σ' is the intersection number of Z_σ with Σ', which is just $\langle c_1(L), \Sigma' \rangle$, by the Corollary in §1.6. In particular, we see that that $[\Phi_{Z_\sigma}] = c_1(L)$ and hence

$$N Z_\sigma = L | Z_\sigma, \tag{3.14}$$

which is sometimes called the *adjunction formula* (see [19], page 146). Thus $\mu([Z_\sigma]) = c_1(L)$, and μ must be onto. Since $H_2(M; \mathbb{Z})$ and $H^2(M; \mathbb{Z})$ are free abelian groups of the same rank, μ is an isomorphism. The fact that μ is an isomorphism is one manifestation of Poincaré duality (for cohomology with integer coefficients).

If β is a smooth two-form on M it follows quickly from Fubini's theorem that

$$\int_M \Phi_\Sigma \wedge \beta = \int_\Sigma i^* \beta,$$

where the i on the right denotes the inclusion of Σ in M. Thus

$$Q(\mu([\Sigma]), [\beta]) = \int_\Sigma i^*\beta = \langle[\Sigma], [\beta]\rangle.$$

Finally, we can use the isomorphism μ to define the intersection form on homology,

$$Q : H_2(M; \mathbb{Z}) \times H_2(M; \mathbb{Z}) \to \mathbb{Z} \quad \text{by} \quad Q([\Sigma], [\Sigma']) = Q(\mu([\Sigma]), \mu([\Sigma'])).$$

We therefore have several equivalent ways of regarding Q:

$$Q([\Sigma], [\Sigma']) = \int_M \Phi_\Sigma \wedge \Phi_{\Sigma'} = \int_\Sigma \Phi_{\Sigma'} = \int_{\Sigma'} \Phi_\Sigma$$

$$= \text{intersection number of } \Sigma \text{ with } \Sigma',$$

when Σ and Σ' intersect transversely. The last interpretation of Q is the motivation for calling it the intersection form.

Since M is simply connected, $H_2(M; \mathbb{Z}) \cong \text{Hom}(H^2(M), \mathbb{Z})$, and hence any group homomorphism $\phi : H^2(M; \mathbb{Z}) \to \mathbb{Z}$ is of the form

$$\phi(b) = \langle[M], b\rangle = Q(\mu([M]), b),$$

for some $[M] \in H_2(M; \mathbb{Z})$. This implies that Q is a *unimodular* symmetric bilinear form. Thus in terms of a basis for the free abelian group $H^2(M; \mathbb{Z})$, the intersection form Q can be represented by a $b_2 \times b_2$ symmetric matrix with integer entries and determinant ± 1.

The integer-valued intersection form Q carries more information than the real-valued form $\tilde{Q}$ defined on de Rham cohomology. Indeed, the integer-valued symmetric bilinear forms represented by the matrices

$$\begin{pmatrix} 0 & 1 \\ 1 & 0 \end{pmatrix}, \quad \begin{pmatrix} 1 & 0 \\ 0 & -1 \end{pmatrix}$$

are equivalent over the reals, but not over the integers because

$$(x \quad y) \begin{pmatrix} 0 & 1 \\ 1 & 0 \end{pmatrix} \begin{pmatrix} x \\ y \end{pmatrix} = 2xy$$

takes on only even values, while

$$(x \quad y) \begin{pmatrix} 1 & 0 \\ 0 & -1 \end{pmatrix} \begin{pmatrix} x \\ y \end{pmatrix} = x^2 - y^2$$

can be either even or odd. in general, we say that a unimodular symmetric bilinear form

$$Q : \mathbb{Z}^r \times \mathbb{Z}^r \to \mathbb{Z}$$

is *even* if $Q(a, a)$ is always even; otherwise it is said to be *odd*.

It is an interesting problem to classify unimodular symmetric bilinear forms of a given rank up to isomorphism. Indefinite unimodular symmetric bilinear forms are classified in [29]; in the odd case, they are direct sums of (± 1)'s, while in the even case, they are direct sums of H's and $E8$'s, where these forms are represented by the matrices

$$H = \begin{pmatrix} 0 & 1 \\ 1 & 0 \end{pmatrix}$$

and

$$E8 = \begin{pmatrix}
-2 & 1 & 0 & 0 & 0 & 0 & 0 & 0 \\
1 & -2 & 1 & 0 & 0 & 0 & 0 & 0 \\
0 & 1 & -2 & 1 & 0 & 0 & 0 & 0 \\
0 & 0 & 1 & -2 & 1 & 0 & 0 & 0 \\
0 & 0 & 0 & 1 & -2 & 1 & 0 & 1 \\
0 & 0 & 0 & 0 & 1 & -2 & 1 & 0 \\
0 & 0 & 0 & 0 & 0 & 1 & -2 & 0 \\
0 & 0 & 0 & 0 & 1 & 0 & 0 & -2
\end{pmatrix}.$$

This last matrix is negative-definite, has only even entries down the diagonal, and has determinant one. It can be shown to be the smallest rank matrix with these properties.

On the other hand, there are often many definite forms of a given rank. For example, in the even case, there are over 10^{51} negative definite forms of rank 40!

Here are the intersection forms of some important simply connected four-manifolds.

Example 1. The intersection form of the complex projective plane $P^2\mathbb{C}$ with its usual orientation is represented by the matrix

$$(1).$$

This follows immediately from the fact that the cohomology ring of $P^2\mathbb{C}$ is generated by an element $a \in H^2(P^2\mathbb{C})$ subject to a single relation $a^3 = 0$. The intersection form of the complex projective plane with the opposite orientation, denoted by $\overline{P^2\mathbb{C}}$, is

$$(-1).$$

Example 2. The intersection form of $S^2 \times S^2$ is represented by the matrix H. Indeed, if ω_1 and ω_2 are the area forms of the two factors, normalized to give area one,

$$\left(\int_{S^2 \times S^2} \omega_i \wedge \omega_j \right) = \begin{pmatrix} 0 & 1 \\ 1 & 0 \end{pmatrix}.$$

Example 3. The K3 surface is the smooth complex hypersurface in $P^3\mathbb{C}$ which is defined in homogeneous coordinates by the equation

$$(z_0)^4 + (z_1)^4 + (z_2)^4 + (z_3)^4 = 0.$$

As we will see in §3.8, this four-manifold has an even intersection form and the topological invariants

$$b_+ = 3, \qquad b_- = 19.$$

It then follows from the classification of indefinite forms that the intersection form of this algebraic surface is

$$E8 \oplus E8 \oplus H \oplus H \oplus H.$$

Example 4. Let M_1 and M_2 be two compact oriented four-manifolds. Choose points $p \in M_1$ and $q \in M_2$ and positively oriented coordinate systems (U, ϕ) and (V, ψ) centered at p and q. By rescaling the coordinates if necessary, we can arrange that

$$B_2(0) \subset \phi(U), \qquad B_2(0) \subset \psi(U),$$

where $B_2(0)$ is the ball of radius two about the origin in $\mathbb{R}^4$. We form the *oriented connected sum* $M_1 \natural M_2$ as the quotient of

$$[M_1 - \phi^{-1}(B_{1/2}(0))] \cup [M_2 - \psi^{-1}(B_{1/2}(0))]$$

by identifying $\phi^{-1}(x)$ with $\psi^{-1}(x/|x|^2)$. This corresponds to taking a disk out of each manifold and identifying the boundaries by an orientation reversing diffeomorphism. If M_1 and M_2 have intersection forms Q_1 and Q_2 respectively, then the oriented connected sum $M_1 \natural M_2$ has intersection form $Q_1 \oplus Q_2$.

An element $a \in H^2(M;\mathbb{Z})$ is said to be *characteristic* if it satisfies the criterion

$$Q(a, b) = Q(b, b) \mod 2, \qquad \text{for all } b \in H^2(M;\mathbb{Z}).$$

Theorem. *If L is the virtual line bundle for a spin^c structure on a smooth four-manifold M, then $c_1(L^2)$ is a characteristic element. Conversely, any characteristic element is $c_1(L^2)$ for some spin^c structure on M.*

Proof: If E is a genuine line bundle on M, then $L \otimes E$ corresponds to a second spin^c structure. Using (2.17), we calculate the difference between the indices of the Dirac operators with coefficients in L and $L \otimes E$, and conclude that

$$\int_M c_1(L) \wedge c_1(E) + \frac{1}{2} \int_M (c_1(E))^2 \in \mathbb{Z},$$

which immediately implies that

$$\frac{1}{2} Q(c_1(L^2), c_1(E)) + \frac{1}{2} Q(c_1(E), c_1(E)) = 0, \quad \mod \mathbb{Z}.$$

Since any element of $H^2(M; \mathbb{Z})$ is the first Chern class of some line bundle E, we conclude that $c_1(L^2)$ is characteristic.

We prove the converse using the the fact that every oriented four-manifold has a spin^c structure. Thus there exists at least one characteristic element a which corresponds to a spin^c structure with some virtual line bundle, which we call L. If a' is a second characteristic element, then

$$Q(a' - a, b) = 0 \quad \mod 2, \qquad \text{for all } b \in H^2(M; \mathbb{Z}),$$

from which it follows that $a' - a$ is divisible by two. Thus there is a line bundle E over M with $c_1(E^2) = a' - a$, and we can construct a spin^c structure on M with virtual line bundle $L \otimes E$. Clearly $c_1((L \otimes E)^2) = a'$.

Corollary. *A smooth four-manifold has an even intersection form if and only if it has a spin structure.*

Proof: Simply note that Q is even if and only if 0 is a characteristic element.

The intersection form is the basic topological invariant of a compact simply connected topological four-manifold. In fact, a classical theorem of J. H. C. Whitehead and John Milnor states that two such four-manifolds are homotopy equivalent if and only if they have the same intersection form. More recently, Michael Freedman showed that any unimodular symmetric bilinear form is realized as the intersection form of a compact simply connected topological four-manifold, and that there are at most two such manifolds with the same intersection form [17].

Not all unimodular symmetric bilinear forms are realized on smooth four-manifolds, however. In particular, although $E8$ can be realized as the intersection form of a compact oriented topological four-manifold, it cannot be realized on a smooth manifold, because such a manifold would be spin by the above theorem and hence would contradict Rochlin's theorem.

3.6   Donaldson's Theorem

There is a very striking restriction on the definite intersection forms that can be realized on a a simply connected four-manifold which possesses a smooth structure:

Theorem (Donaldson [10]). *The only negative definite unimodular form represented by a compact smooth simply connected four-manifold is* $Q = -I$.

There are two cases to the proof, Q even and Q odd. We will only give a complete proof in the even case—we will show that an even negative definite form (such as $E8 \oplus E8$) cannot be realized on a smooth four-manifold.

In the even case, we can take L to be the trivial bundle and use Transversality Theorem 1 of §3.4. Since $b_+ = 0$, this theorem implies that $\widetilde{\mathcal{M}}_\phi$ is a smooth manifold of dimension $k = b_-/4$, and the quotient manifold $\mathcal{M}_\phi$ is a smooth manifold of dimension $k - 1$ except for a singularity at the point $[d_A, 0]$, d_A being the unique connection for which $F_A^+ = \phi$. The tangent space to $\widetilde{\mathcal{M}}_\phi$ at d_A is simply the space of harmonic spinor fields $\ker(D_A^+)$ on which the group S^1 of unit-length complex numbers acts in the usual fashion by complex multiplication. Hence $\mathcal{M}_\phi$ is a smooth manifold except for a singularity at $[d_A, 0]$ which is a cone on $P^{m-1}\mathbb{C}$, where $2m = k$.

Now we recall the fiber bundle (3.8) in which the total space $\widetilde{B}^*$ is the bundle of unit-length vectors in the universal bundle E. We have a commutative diagram

$$
\begin{array}{ccccc}
\widetilde{\mathcal{M}}_\phi - \{[d_A, 0]\} & \subset & \widetilde{B}^* & = & E \\
\downarrow & & \downarrow & & \downarrow \\
\mathcal{M}_\phi - \{[d_A, 0]\} & \subset & B^* & = & P^\infty\mathbb{C}
\end{array}
$$

The bundle E over $P^\infty\mathbb{C}$ restricts to a bundle over $P^{m-1}\mathbb{C}$ which we easily verify to be the Hopf fibration

$$
S^1 \to S^{2m-1} \to P^{m-1}\mathbb{C}.
$$

Hence the restriction of E to $P^{m-1}\mathbb{C}$ has nontrivial first Chern class, and since the cohomology of $P^{m-1}\mathbb{C}$ is a truncated polynomial algebra,

$$
H^*(P^{m-1}\mathbb{C}; \mathbb{Z}) \cong P[a]/(a^m),
$$

where a has degree two, we see that

$$
\int_{P^{m-1}\mathbb{C}} (c_1(E))^{m-1} \neq 0 \qquad \Rightarrow \qquad [P^{m-1}\mathbb{C}] \neq 0
$$

as an element of $H_{2m-2}(\mathcal{B}^*; \mathbb{Z})$. But this contradicts the fact that $\mathcal{M}_\phi$ exhibits $P^{m-1}\mathbb{C}$ as a boundary in $\mathcal{B}^*$.

Here is a sketch of how to proceed if in the case where Q is odd. In this case, the argument in the preceding paragraphs will yield a contradiction so long as a characteristic element can be found such that the corresponding moduli space $\mathcal{M}_\phi$ has positive formal dimension.

Since $b_+ = b_1 = 0$, this will occur exactly when the index of D_A^+ is positive. By the Atiyah-Singer index theorem, this means that

$$-\frac{\tau(M)}{8} + \frac{1}{2}\int_M c_1(L)^2 = \frac{1}{8}(b_- + Q(a,a)) > 0, \qquad \text{or} \qquad -Q(a,a) < b_2,$$

since $b_2 = b_-$. An argument of Elkies [15] shows that either $Q = -I$ or there is a characteristic element satisfying this last inequality, thereby completing this approach to proving Donaldson's theorem.

Donaldson's theorem gives a complete answer to the question of which definite unimodular symmetric bilinear forms can be realized as intersection forms on compact simply connected smooth four-manifolds—the only possibilities are the intersection forms of

$$P^2\mathbb{C} \,\natural\, \cdots \,\natural\, P^2\mathbb{C} \qquad \text{and} \qquad \overline{P^2\mathbb{C}} \,\natural\, \cdots \,\natural\, \overline{P^2\mathbb{C}}.$$

One could ask a similar question for the indefinite forms. All the odd indefinite forms can be realized on connected sums of $P^2\mathbb{C}$'s and $\overline{P^2\mathbb{C}}$'s. If $s \geq 3r$, the even form

$$2r(E8) \oplus sH$$

can be realized on a connected sum of r copies of the $K3$ surfaces and $s - 3r$ copies of $S^2 \times S^2$. One of the striking successes of the new Seiberg-Witten theory is a theorem due to Furuta: If $2rE8 \oplus sH$ is realized on a compact simply connected smooth four-manifold, then $s \geq 2r+1$. (See [1] for further discussion.)

Coupled with results of Freedman, which are described in [17], Donaldson's Theorem implies that there are smooth four-manifolds which are homeomorphic but not diffeomorphic to $\mathbb{R}^4$. A relatively elementary presentation of the basic idea is presented in [16], §1.

3.7   Seiberg-Witten invariants

For four-manifolds with indefinite intersection form, the Seiberg-Witten equations provide a collection of invariants which depend upon the choice of a spin^c structure. These invariants sometimes allow us to distinguish between smooth structures on a given topological four-manifold.

Let L be a virtual complex line bundle over M such that $W_+ \otimes L$ is defined as a genuine complex vector bundle. Assuming that $b_+(M) > 0$, we can then form the moduli space $\mathcal{M}_{L,\phi}$, which depends upon L. According to the Transversality Theorem 2 from §3.4, this moduli space is a compact manifold for a generic choice of ϕ. The key point in the argument is that a solution to the perturbed Seiberg-Witten equations with $\psi = 0$ can occur only if $c_1(L)$ contains a connection with $F_A^+ = 0$. This will not occur generically, but only for ϕ lying in an affine subspace Π of $\Omega_+^2(M)$ of codimension $b_+ >$ one, by the Proposition at the end of §1.9.

Of course, the moduli space $\mathcal{M}_{L,\phi}$ depends upon the choice of Riemannian metric on M and the choice of ϕ. If ϕ_1 and ϕ_2 are two choices and $b_+ \geq 2$, then a generic path γ from ϕ_1 to ϕ_2 will miss the subspace Π. It follows from Smale's infinite-dimensional generalization of transversality (see [36], Theorems 3.1 and 3.3) that $\mathcal{W} = \pi^{-1}(\gamma)$ is an oriented submanifold of $\mathcal{N}$ such that

$$\partial \mathcal{W} = \mathcal{M}_{L,\phi_2} - \mathcal{M}_{L,\phi_1}.$$

In other words, different choices of ϕ yield cobordant moduli spaces. It follows that if $[\alpha]$ is any cohomology class in $\mathcal{B}^*$, then

$$\langle [\alpha], [\mathcal{M}_{L,\phi_1}] \rangle = \langle [\alpha], [\mathcal{M}_{L,\phi_2}] \rangle. \tag{3.15}$$

Similarly, changing the Riemannian metric on M alters the moduli space by a cobordism.

Definition. Suppose that $b_+ \geq 2$ and that the dimension of $\mathcal{M}_{L,\phi}$ is even. Then the *Seiberg-Witten invariant* of L is

$$\mathrm{SW}(L) = \langle c_1^d, [\mathcal{M}_{L,\phi}] \rangle, \quad \text{where} \quad d = \frac{1}{2} \dim \mathcal{M}_{L,\phi},$$

c_1 is the first Chern class of the complex line bundle E associated to the S^1-bundle $\widetilde{\mathcal{B}}^* \to \mathcal{B}^*$, and ϕ is chosen to be generic.

Equation (3.15) and the corresponding equation for change of Riemannian metrics shows that the Seiberg-Witten invariants are well-defined.

If $b_+ = 1$, we can still define $SW(L)$ when $\langle c_1(L)^2, [M] \rangle \geq 0$ and $L \neq 0$. These conditions imply that L does not admit a connection with $F_A^+ = 0$, and hence no connection with $F_A^+ = \phi$, when ϕ is sufficiently small, so we can take

$$\mathrm{SW}(L) = \langle c_1^d, [\mathcal{M}_{L,\phi}] \rangle, \quad \text{for generic small choice of } \phi.$$

Theorem (Witten [41]). *If an oriented Riemannian manifold has a Riemannian metric of positive scalar curvature, all of its Seiberg-Witten invariants vanish.*

Proof: An immediate consequence of the Lemma from §3.3.

Manifolds which admit metrics of positive scalar curvature include S^4, $P^2\mathbb{C}$ and $\overline{P^2\mathbb{C}}$. A theorem of Gromov and Lawson [20] implies that if M and N are four-manifolds of positive scalar curvature, their connected sum $M \natural N$ also admits a metric of positive scalar curvature. Thus Witten's theorem implies that the manifolds

$$k(P^2\mathbb{C}) \,\natural\, l(\overline{P^2\mathbb{C}})$$

all have vanishing Seiberg-Witten invariants.

The simplest examples of manifolds with nonvanishing Seiberg-Witten invariants come from the theory of complex surfaces, as we will see in the next two sections. In these cases, the moduli space consists of a finite number of points, with signs.

The formal dimension of the moduli space $\mathcal{M}_L$ is zero when

$$c_1(L^2)^2[M] = 3\tau(M) + 2\chi(M),$$

which according to (2.20) occurs exactly when $c_2(W_+ \otimes L) = 0$. The following lemma explains the relationship between this last condition and the existence of almost complex structures.

Lemma. *Homotopy classes of almost complex structures on M are in one-to-one correspondence with line bundles L^2 over M such that*

1. *$W_+ \otimes L$ exists as a bundle,*

2. *$c_1(L^2)^2[M] = 3\tau(M) + 2\chi(M)$.*

Proof: An almost complex structure J on M yields a spin^c-structure as we described in §2.3. If L is the corresponding virtual line bundle, then $W_+ \otimes L$ exists as a genuine vector bundle and divides as a Whitney sum,

$$W_+ \otimes L = \Theta \oplus L^2,$$

where Θ is the trivial line bundle. It follows from Proposition 4 from §1.5 that

$$\mathrm{ch}(W_+ \otimes L) = 1 + \mathrm{ch}(L^2) = 2 + c_1(L^2) + \frac{1}{2}(c_1(L^2))^2.$$

Thus $c_1(W_+ \otimes L) = c_1(L^2)$, and by (2.18), $c_2(W_+ \otimes L) = 0$. The second condition of the Lemma now follows from (2.20).

To prove the converse, note that if the conditions in the statement of the Lemma hold, then according to (2.20), $c_2(W_+ \otimes L) = 0$. It then follows from the Classification Theorem for Quaternionic Line Bundles from §1.2 that $W_+ \otimes L$ is trivial. Thus $W_+ \otimes L$ has a nowhere vanishing section ψ, and

$$\omega = \frac{\sigma(\psi)}{|\sigma(\psi)|}$$

is a unit-length self-dual two-form on M. We can therefore define an almost complex structure $J(L^2)$ on M by

$$\omega(Jx, \cdot) = -\langle x, \cdot \rangle.$$

It is readily checked that the two construction we have described are inverse to each other.

3.8 Dirac operators on Kähler surfaces

Among the most important examples of smooth four-manifolds are Kähler surfaces, Kähler manifolds of complex dimension two. In fact, it is relatively easy to do computations with these manifolds.

Recall that a Riemannian metric $\langle \, , \, \rangle$ on a complex manifold M is called *Hermitian* if

$$\langle Jx, Jy \rangle = \langle x, y \rangle, \qquad \text{for all} \quad x, y \in T_pM.$$

In this case we can define a rank two covariant tensor field ω on M by

$$\omega(x, y) = \langle Jx, y \rangle. \tag{3.16}$$

Since

$$\omega(y, x) = \langle Jy, x \rangle = \langle J^2y, Jx \rangle = -\langle y, Jx \rangle = -\langle Jx, y \rangle = -\omega(x, y),$$

we see that ω is skew-symmetric. Thus ω is a two-form, called the *Kähler form* on M.

Definition. A *Kähler manifold* is a complex manifold M with a Hermitian metric $\langle \, , \, \rangle$ whose Kähler form is closed.

In terms of a moving orthonormal frame (e_1, e_2, e_3, e_4) such that $e_2 = Je_1$ and $e_4 = Je_3$, we can express the Kähler form as

$$\omega = e_1 \wedge e_2 + e_3 \wedge e_4.$$

This formula makes it clear that the Kähler form ω is self-dual. The Kähler condition $d\omega = 0$ therefore implies also that $\delta\omega = 0$, so $\omega \in \mathcal{H}_+^2(M)$.

Example 1. The complex projective space $P^N\mathbb{C}$ can be made into a Kähler manifold. To do this, we let $\pi : \mathbb{C}^{N+1} - \{0\} \to P^N\mathbb{C}$ denote the usual projection. If U is an open subset of $P^N\mathbb{C}$ and $Z : U \to \mathbb{C}^{N+1} - \{0\}$ is a holomorphic map with $\pi \circ Z = \text{id}$, we set

$$\omega = i\partial\bar{\partial}\log(|Z|^2) = i\sum_{i,j=1}^N \frac{\partial^2}{\partial z_i\partial\bar{z}_j}\log(|Z|^2)dz_id\bar{z}_j.$$

It is readily checked that ω is independent of the choice of holomorphic section Z: if $Z = fW$, where f is a nonzero holomorphic complex-valued function on U, then

$$i\partial\bar{\partial}\log(|Z|^2) = i\partial\bar{\partial}[\log(|f|^2) + \log(|W|^2)]$$
$$= i\partial\bar{\partial}\log f + i\bar{\partial}\partial\log\bar{f} + i\partial\bar{\partial}\log(|W|^2) = i\partial\bar{\partial}\log(|W|^2).$$

Moreover, ω is closed and a calculation in local coordinates shows that it is of maximal rank ($\omega^N \neq 0$). One uses (3.16) in reverse to define a Hermitian metric on $P^N\mathbb{C}$ which makes $P^N\mathbb{C}$ into a Kähler manifold with the Kähler form ω.

Example 2. Any complex submanifold of a Kähler manifold is itself a Kähler manifold in the induced metric. According to a famous theorem of Chow, compact complex submanifolds of $P^N\mathbb{C}$ are projective algebraic varieties without singularities, each describable as the zero locus of a finite collection of homogeneous polynomials.

For example, given a positive integer n, we can consider the complex surface $M_n \subset P^3\mathbb{C}$ defined in homogeneous coordinates by the polynomial equation

$$P(z_0, z_1, z_2, z_3) = (z_0)^n + (z_1)^n + (z_2)^n + (z_3)^n = 0. \qquad (3.17)$$

It follows from a theorem of Lefschetz on hyperplane sections (which is proven on pages 156-159 of [19] and in §7 of [28]) that M_n is simply connected. The other topological invariants of M_n can be calculated by regarding the polynomial P as a section of H^n, where H is the hyperplane section bundle over $P^3\mathbb{C}$.

The *hyperplane section bundle* H over $P^N\mathbb{C}$ is defined via the open covering $\{U_0, \ldots U_N\}$, where

$$U_i = \{[z_1, \ldots, z_N] \in P^N\mathbb{C} : z_i \neq 0\},$$

by the transition functions $g_{ij} = (z_j/z_i)$. A linear form $a_0z_0 + \cdots + a_Nz_N$ defines a section of H with local representatives

$$\sigma_i = a_0\frac{z_0}{z_i} + \cdots + a_N\frac{z_N}{z_i},$$

and similarly, any polynomial of degree n (such as P) will define a section of H^n. By the adjunction formula (3.14) the normal bundle to M_n is simply the restriction of H^n to M_n.

The hyperplane section bundle is the inverse of the universal bundle, which has total space E_∞ given by (1.21), as we described in §1.7. Recall that the first Chern class of E_∞ is $-a$, where a is the standard generator of $H^2(P^N\mathbb{C}; \mathbb{Z})$ such that $\langle [P^1\mathbb{C}], a \rangle = 1$. (Equivalently, a is the cohomology class of the Kähler form ω.) Thus the first Chern class of H is a, and the Chern characters of these bundles are

$$\mathrm{ch}(E_\infty) = e^{-a}, \qquad \mathrm{ch}(H) = e^a.$$

Using Proposition 4 of §1.5, we see that

$$\mathrm{ch}(H^n) = e^{na}, \qquad \mathrm{ch}(E_\infty^\perp) = (N+1) - e^{-a},$$

the latter equality following from the fact that $E_\infty \oplus E_\infty^\perp$ is the trivial bundle of rank $N + 1$.

On the other hand, we have an isomorphism

$$TP^N\mathbb{C} \cong \mathrm{Hom}(E_\infty, E_\infty^\perp) = E_\infty^* \otimes E_\infty^\perp = L \otimes E_\infty^\perp;$$

to see this, note that complex lines in $\mathbb{C}^{N+1}$ near a given complex line $[z_0, \ldots, z_N]$ correspond to linear maps from the one-dimensional linear space spanned by $(z_0, \ldots, z_N)$ to its orthogonal complement. It follows that

$$\mathrm{ch}(TP^N\mathbb{C}) = \mathrm{ch}(L)\mathrm{ch}(E_\infty^\perp) = e^a[(N+1) - e^{-a}] = (N+1)e^a - 1.$$

We can apply these calculations to our hypersurface $M = M_n \subset P^3\mathbb{C}$. Since

$$TP^3\mathbb{C} = TM \oplus NM \cong TM \oplus (H^n)|M,$$

we see that

$$4e^a - 1 = \mathrm{ch}(TM) + e^{na} \quad \text{or} \quad \mathrm{ch}(TM) = 4e^a - e^{na} - 1,$$

where a now denotes the pullback of the standard generator of $H^2(P^N\mathbb{C}; \mathbb{Z})$ to M. Hence

$$3 + c_1(TM) + \frac{1}{2}[c_1(TM)]^2 - c_2(TM) = 3 + (4-n)a + \frac{1}{2}(4-n^2)a^2,$$

modulo terms of degree greater than four, from which we conclude that

$$c_1(TM) = (4-n)a, \qquad c_2(TM) = (n^2 - 4n + 6)a^2.$$

As explained in detail in [19], pages 171 and 172, the Kähler surface M_n intersects a generic line in $P^3\mathbb{C}$ in n points and therefore has degree n in the sense of algebraic geometry. By a theorem of Wirtinger, its volume is therefore n times as large as that of $P^2\mathbb{C}$, and hence $\langle a, [M_n]\rangle = n$. Therefore,

$$\langle [c_1(TM)]^2, [M]\rangle = (4-n)n, \qquad \langle c_2(TM), [M]\rangle = n^3 - 4n^2 + 6n.$$

In particular, in the case $n = 4$, which gives the K3 surface described in §3.5,

$$c_1(TM) = 0 \quad \text{and} \quad \langle c_2(TM), [M]\rangle = 24.$$

But we will see shortly (3.22) that $c_2(TM)[M] = \chi(M)$, the Euler characteristic of M, and hence

$$b_0 = b_4 = 1 \quad \text{and} \quad b_1 = b_3 = 0 \qquad \Rightarrow \qquad b_2 = 22.$$

Since M has a complex structure, it follows from the Lemma at the end of the previous section that

$$3\tau(M) + 2\chi(M) = 0, \quad \text{which implies} \quad \tau(M) = -16,$$

from which it follows that

$$b_+ = 3 \quad \text{and} \quad b_- = 19,$$

as we claimed in §3.5. Moreover, the complex structure defines a canonical spin^c structure on M with virtual bundle L satisfying $c_1(L^2) = 0$. Thus L must be trivial, M_4 must have a genuine spin structure and its intersection form must be even.

A similar computation in the case $n = 5$ would yield

$$b_+ = 9 \quad \text{and} \quad b_- = 44;$$

in this case, $c_1(L^2)$ is odd and hence 0 is not characteristic, so the intersection form is odd.

On a complex manifold M, a (p,q)-form is a complex-valued differential form which can be expressed in local complex coordinates $(z_1, \ldots, z_n)$ as

$$\sum f_{i_1 \cdots i_p, j_1 \cdots j_q} dz_{i_1} \wedge \cdots \wedge dz_{i_p} \wedge d\bar{z}_{j_1} \wedge \cdots \wedge d\bar{z}_{j_q}.$$

We can decompose the space of complex-valued k-forms into forms of type (p,q),

$$\Omega^k(M) \otimes \mathbb{C} = \sum_{p+q=k} \Omega^{p,q}(M), \qquad \Omega^{p,q}(M) = \{(p,q)\text{-forms}\}. \qquad (3.18)$$

The exterior derivative divides into two components,

$$d = \partial + \bar{\partial}, \quad \partial : \Omega^{p,q}(M) \to \Omega^{p+1,q}(M), \quad \bar{\partial} : \Omega^{p,q}(M) \to \Omega^{p,q+1}(M),$$

and the Kähler condition implies that $\partial \circ \partial = 0$ and $\bar{\partial} \circ \bar{\partial} = 0$.

The Hermitian metric $\langle \cdot, \cdot \rangle$ on TM extends to $TM \otimes \mathbb{C}$ so that the extension is complex linear in the first variable and conjugate linear in the second. This in turn defines a Hermitian metric on each $\Lambda^k T^* M \otimes \mathbb{C}$ and a Hermitian inner product $(\cdot, \cdot)$ on the space of smooth k-forms by

$$(\theta, \phi) = \int_M \langle \theta, \phi \rangle (1/n!) \omega^n,$$

where ω is the Kähler form. Note that the decomposition (3.18) is orthogonal with respect to this inner product. We can therefore define formal adjoints,

$$\partial^* : \Omega^{p+1,q}(M) \to \Omega^{p,q}(M), \quad \bar{\partial}^* : \Omega^{p,q+1}(M) \to \Omega^{p,q}(M),$$

just as in §1.9, by requiring

$$(\partial^* \theta, \phi) = (\theta, \partial \phi), \qquad (\bar{\partial}^* \theta, \phi) = (\theta, \bar{\partial} \phi). \tag{3.19}$$

In the case of complex surfaces, the Hermitian metric reduces the structure group to $U(2)$, and the various representations of $U(2)$ correspond to the different types of tensor fields, sections of various associated vector bundles. For example, the standard representation of $U(2)$ on C^2 yields the complex tangent bundle TM, while the representation $A \mapsto {}^t A^{-1} = \bar{A}$ yields the complex cotangent bundle whose sections are the $(1,0)$-forms on M. If the transition functions for TM are

$$g_{\alpha\beta} : U_\alpha \cap U_\beta \to U(2),$$

then the transition functions for the cotangent bundle are

$${}^t g_{\alpha\beta}^{-1} = \bar{g}_{\alpha\beta} : U_\alpha \cap U_\beta \to U(2).$$

Sections of the conjugate of this bundle are either $(0,1)$-forms or complex vector fields—in the presence of a Hermitian metric, we can identify $(0,1)$-forms with complex vector fields.

The complex exterior power $K = \Lambda^2_\mathbb{C} T^* M$ of the cotangent bundle is called the *canonical bundle*. Its sections are $(2,0)$-forms on M and it has transition functions

$$(\det \circ {}^t g_{\alpha\beta}^{-1}) = \det \circ \bar{g}_{\alpha\beta} : U_\alpha \cap U_\beta \to U(1).$$

The canonical bundle plays a key role in the celebrated classification of complex surfaces by Enriques and Kodaira (see [22] or [5]). Conjugation gives the anti-canonical bundle whose sections are the $(0,2)$-forms and whose transition functions are

$$\det \circ g_{\alpha\beta} : U_\alpha \cap U_\beta \to U(1).$$

We need to relate all of this to the canonical $\mathrm{Spin}(4)^c$-structure constructed in §2.3. Recall that the transition functions for the distinguished line bundle L^2 selected by the $\mathrm{Spin}(4)^c$-structure are

$$\det \circ g_{\alpha\beta} : U_\alpha \cap U_\beta \to U(1),$$

so L^2 is just the anti-canonical bundle. Moreover,

$$W_+ \otimes L \cong W_+ \otimes K^{-(1/2)} \cong \Theta \oplus K^{-1}, \tag{3.20}$$

where Θ denotes the trivial line bundle. On the other hand, since $\rho^c \circ j$ is the usual representation of $U(2)$ on $\mathbb{C}^2$,

$$W_- \otimes L \cong TM, \tag{3.21}$$

the tangent bundle of M with its almost complex structure. Observe that it follows from (3.20) that $c_2(W_+ \otimes L) = 0$, and then from (2.20) and (2.21) that $c_2(W_- \otimes L)[M] = \chi(M)$, the Euler characteristic of M. Hence, by (3.21),

$$c_2(TM)[M] = \chi(M). \tag{3.22}$$

By the method given in §2.5, we can constuct a $\mathrm{Spin}(4)^c$ connection on $W \otimes L$, with local representative

$$d - iaI - \frac{1}{4} \sum_{i,j=1}^{4} \omega_{ij} e_i \cdot e_j, \tag{3.23}$$

the local moving orthonormal frame (e_1, e_2, e_3, e_4) being chosen so that the Kähler form is

$$e_1 \cdot e_2 + e_3 \cdot e_4 = \begin{pmatrix} -2i & 0 & 0 & 0 \\ 0 & 2i & 0 & 0 \\ 0 & 0 & 0 & 0 \\ 0 & 0 & 0 & 0 \end{pmatrix}.$$

The local representative (3.23) can be rewritten in matrix form as

$$d - iaI - \frac{1}{2} \begin{pmatrix} -i(\omega_{12} + \omega_{34}) & (\omega_{13} + \omega_{42}) - i(\omega_{14} + \omega_{23}) & 0 & 0 \\ \sharp & i(\omega_{12} + \omega_{34}) & 0 & 0 \\ 0 & 0 & \sharp & \sharp \\ 0 & 0 & \sharp & \sharp \end{pmatrix}.$$

Since the Kähler form is parallel with repect to the Levi-Civita connection,

$$\omega_{13} + \omega_{42} = 0 \qquad \text{and} \qquad \omega_{14} + \omega_{23} = 0.$$

We determine a connection d_{A_0} on $K^{-(1/2)}$ by setting

$$a = \frac{1}{2}(\omega_{12} + \omega_{34}),$$

so that the induced connection on the trivial summand Θ of $W_+ \otimes L$ is the trivial connection. The connection induced on the other summand K^{-1} is then exactly the same as that induced by the Levi-Civita connection.

Proposition. *In the Kähler case, the Dirac operator corresponding to the* Spin(4)^c *connection d_{A_0} is*

$$D_{A_0} = \sqrt{2}(\bar{\partial} + \bar{\partial}^*), \qquad (3.24)$$

when we identify sections of Θ with $(0,0)$-forms, sections of K^{-1} with $(0,2)$-forms, and sections of $W_- \otimes L = TM$ with $(0,1)$-forms.

Sketch of proof: It suffices to show that the two sides of the above equation agree at a given point of M. A key feature of Kähler manifolds (discussed in [19], Chapter 1, §7) is that one can choose local coordinates $z_1 = x_1 + ix_2$, $z_2 = x_3 + ix_4$, which are "normal" at p, so that

$$g_{ij}(p) = \delta_{ij}, \qquad (\partial g_{ij}/\partial x_k)(p) = 0.$$

This implies that covariant derivatives can be replaced by ordinary derivatives of components at p. We set $e_i(p) = (\partial/\partial x_i)(p)$ and choose

$$\epsilon_1(p), \epsilon_2(p) \in (W_+ \otimes L)_p, \qquad \epsilon_3(p), \epsilon_4(p) \in (W_- \otimes L)_p$$

with respect to which $(e_1(p), \ldots, e_4(p))$ are represented by the matrices given in §2.4. By means of the isomorphisms (3.20) and (3.21), we can make the identifications

$$\epsilon_1(p) = 1, \qquad \epsilon_2(p) = \frac{1}{2}(d\bar{z}_2 \wedge d\bar{z}_1)(p),$$

$$\epsilon_3(p) = \frac{1}{\sqrt{2}}d\bar{z}_1(p), \qquad \epsilon_4(p) = \frac{1}{\sqrt{2}}d\bar{z}_2(p),$$

in which the $(1/\sqrt{2})$'s occur to normalize $d\bar{z}_1(p)$ and $d\bar{z}_2(p)$ to unit length.

At the point p, the Dirac operator can be expressed in matrix form as

$$\begin{pmatrix} 0 & 0 & -\frac{\partial}{\partial x_1} + i\frac{\partial}{\partial x_2} & -\frac{\partial}{\partial x_3} + i\frac{\partial}{\partial x_4} \\ 0 & 0 & \frac{\partial}{\partial x_3} + i\frac{\partial}{\partial x_4} & -\frac{\partial}{\partial x_1} - i\frac{\partial}{\partial x_2} \\ \frac{\partial}{\partial x_1} + i\frac{\partial}{\partial x_2} & -\frac{\partial}{\partial x_3} + i\frac{\partial}{\partial x_4} & 0 & 0 \\ \frac{\partial}{\partial x_3} + i\frac{\partial}{\partial x_4} & \frac{\partial}{\partial x_1} - i\frac{\partial}{\partial x_2} & 0 & 0 \end{pmatrix}.$$

Applying our identifications, we find that

$$D_{A_0}\begin{pmatrix} f \\ 0 \\ 0 \\ 0 \end{pmatrix} = \begin{pmatrix} 0 \\ 0 \\ \frac{\partial f}{\partial x_1} + i\frac{\partial f}{\partial x_2} \\ \frac{\partial f}{\partial x_3} + i\frac{\partial f}{\partial x_4} \end{pmatrix} = \sqrt{2}\left(\frac{\partial f}{\partial \bar{z}_1}d\bar{z}_1 + \frac{\partial f}{\partial \bar{z}_2}d\bar{z}_2\right),$$

which is just $\sqrt{2}(\bar{\partial}f)(p)$. Similarly,

$$D_{A_0}\begin{pmatrix} 0 \\ 0 \\ h_1 \\ h_2 \end{pmatrix} = \begin{pmatrix} \frac{\partial h_1}{\partial x_3} + i\frac{\partial h_1}{\partial x_4} - \frac{\partial h_2}{\partial x_1} - i\frac{\partial h_2}{\partial x_2} \\ 0 \\ 0 \end{pmatrix}^{\#}.$$

Thus

   the K^{-1}-component of $D_{A_0}(h_1 d\bar{z}_1 + h_2 d\bar{z}_2) = \sqrt{2}\bar{\partial}(h_1 d\bar{z}_1 + h_2 d\bar{z}_2)$.

If $\Pi^{(0,2)}$ denotes the orthogonal projection on $(0,2)$-forms, then

$$D_{A_0} = \sqrt{2}\bar{\partial} : \Omega^{(0,0)}(M) \longrightarrow \Omega^{(0,1)}(M),$$

$$\Pi^{(0,2)} \circ D_{A_0} = \sqrt{2}\bar{\partial} : \Omega^{(0,1)}(M) \longrightarrow \Omega^{(0,2)}(M).$$

The argument for the Proposition is completed by noting that the only self-adjoint operator with these two properties is $\sqrt{2}(\bar{\partial} + \bar{\partial}^*)$.

Remark: Any other Spin(4)^c connection is of the form

$$d_A = d_{A_0} - ia,$$

for some real-valued one-form a on M. In the general case, the above proof shows that

$$D_A^+\begin{pmatrix} f \\ 0 \\ 0 \\ 0 \end{pmatrix} = \begin{pmatrix} 0 \\ 0 \\ \frac{\partial f}{\partial x_1} - ia(\frac{\partial}{\partial x_1})f + i\frac{\partial f}{\partial x_2} + a(\frac{\partial}{\partial x_2})f \\ \frac{\partial f}{\partial x_3} - ia(\frac{\partial}{\partial x_3})f + i\frac{\partial f}{\partial x_4} + a(\frac{\partial}{\partial x_4})f \end{pmatrix} = \sqrt{2}[\bar{\partial}f - (ia)_{(0,1)} \cdot f],$$

$$\tag{3.25}$$

where $(ia)_{(0,1)}$ denotes the $(0,1)$-component of ia.

In the case of the canonical Spin(4)^c-structure associated to a Kähler surface, the Atiyah-Singer Index Theorem for the operator $D_{A_0}^+$ yields a cornerstone for the theory of complex surfaces:

Riemann-Roch Theorem for Kähler surfaces. *The index of* $D^+_{A_0}$ *is given by Noether's formula*

$$\text{complex index of } D^+_{A_0} = \frac{1}{12}\{c_1^2(TM)[M] + c_2(TM)[M]\}.$$

Proof: According to the Index Theorem,

$$\text{index of } D^+_{A_0} = -\frac{1}{8}\tau(M) + \frac{1}{2}c_1(K^{-1})^2[M].$$

By the lemma from the preceding section,

$$-3\tau(M) = 2\chi(M) - c_1(K^{-1})^2[M],$$

and hence

$$\text{index of } D^+_{A_0} = \frac{1}{12}\chi(M) + \frac{1}{12}c_1(TM)^2[M].$$

The formula we want now follows from (3.22).

On the other hand, the index of $\bar{\partial} + \bar{\partial}^*$ can also be calculated directly by means of Hodge theory. We give only a brief indication of how this goes, and refer the reader to [19] for a more complete discussion.

On a Kähler manifold, J and $\star$ are parallel with respect to the Levi-Civita connection, from which it follows that the Hodge Laplacian commutes with orthogonal projection onto the space of (p,q)-forms, and we can therefore write

$$\mathcal{H}^k(M) = \sum_{p+q=k} \mathcal{H}^{p,q}(M),$$

where $\mathcal{H}^{p,q}(M)$ is the space of harmonic (p,q)-forms. Moreover, the Hodge Laplacian can be expressed in terms of the operator $\bar{\partial}$ and its adjoint:

$$\Delta = 2(\bar{\partial} + \bar{\partial}^*)^2.$$

In the case of a Kähler surface, it follows that

$$\text{Index of } (\bar{\partial} + \bar{\partial}^*) = \dim \mathcal{H}^{0,0}(M) + \dim \mathcal{H}^{0,2}(M) - \dim \mathcal{H}^{0,1}(M). \quad (3.26)$$

Note that under conjugation, we have $\mathcal{H}^{0,2}(M) \cong \mathcal{H}^{2,0}(M)$.

Since $d = \bar{\partial}$ on $\Omega^{2,0}(M)$ and $(0,2)$-forms are automatically self-dual, elements of $\mathcal{H}^{2,0}(M)$ are just the $\bar{\partial}$-closed $(2,0)$-forms. These are just the holomorphic sections of the canonical bundle K, and the dimension of the space of such sections is called the *geometric genus* of M. We denote the geometric genus by p_g. If M is simply connected, $\mathcal{H}^{0,1}(M) = 0$ and it follows from (3.26) that

$$\text{Index of } (\bar{\partial} + \bar{\partial}^*) = 1 + p_g.$$

Noether's formula now yields

$$1 + p_g = \frac{1}{12}\{c_1^2(TM)[M] + c_2(TM)[M]\}.$$

The invariants on the right-hand side can be reexpressed in terms of Euler characteristic and signature,

$$1 + p_g = \frac{1}{4}\{\tau(M) + \chi(M)\} = \frac{1}{2}(b_+ + 1),$$

so $b_+ = 2p_g + 1$. In particular, b_+ is odd for any Kähler surface. Holomorphic sections of K or antiholomorphic sections of K^{-1} are self-dual harmonic forms of type $(2,0)$ or $(0,2)$, and these fill out a linear space of dimension $2p_g$. We therefore conclude that the space of self-dual harmonic forms of type $(1,1)$ is exactly one-dimensional, and any self-dual harmonic two-form of type $(1,1)$ must be a constant multiple of the Kähler form.

3.9   Invariants of Kähler surfaces

In this section, we will calculate the Seiberg-Witten invariant $SW(L)$ in the case where M is a Kähler manifold with canonical bundle K and $L = K^{-1/2}$. Recall that in this case, $W_+ = \Theta \oplus K^{-1}$, where Θ is the trivial line bundle.

Theorem (Witten [41]). *A Kähler manifold with $b_1 = 0$ and $b^+ \geq 2$ must satisfy $SW(K^{-1/2}) = \pm 1$.*

Proof: Let A_0 denote the connection on L defined by the Kähler structure, as described in the previous section. Following an idea of Taubes [38], we set $\phi = F_{A_0}^+ + \omega$ in the modified Seiberg-Witten equations so that they become

$$D_A^+\psi = 0, \qquad F_A^+ = \sigma(\psi) + F_{A_0}^+ + \omega. \qquad (3.27)$$

The advantage to these equations is that they have an obvious solution

$$A = A_0, \qquad \psi = \begin{pmatrix} \sqrt{2} \\ 0 \end{pmatrix}.$$

We will prove that the obvious solution is the only solution in the moduli space and that it is nondegenerate. This will show that $SW(L) = \pm 1$.

Suppose that (A, ψ) is any solution to (3.27) and write

$$\psi = \begin{pmatrix} \alpha \\ \beta \end{pmatrix}, \qquad \text{where} \quad \alpha \in \Gamma(\Theta), \quad \beta \in \Gamma(K^{-1}).$$

We make use of a transformation due to Witten,

$$A \mapsto A, \qquad \alpha \mapsto \alpha, \qquad \beta \mapsto -\beta. \tag{3.28}$$

We claim that this transformation leaves invariant the zeros of the functional

$$S_\phi(A, \psi) = \int_M \left[|D_A \psi|^2 + |F_A^+ - \sigma(\psi) - \phi|^2 \right] dV,$$

and hence takes solutions of the perturbed Seiberg-Witten equations to other solutions.

To see this, we write $\nabla^A = \nabla^{A_0} - ia$, where a is a real-valued one-form, so that

$$F_A^+ - \phi = (da)^+ - \omega,$$

and apply the Weitzenböck formula as in §3.1 to obtain

$$S_\phi(A, \psi) = \int_M \left[|\nabla^A \psi|^2 + \frac{s}{4} |\psi|^2 + 2\langle F_A^+, \sigma(\psi) \rangle + |(da)^+ - \sigma(\psi) - \omega|^2 \right] dV.$$

Note that the connection ∇^A preserves the direct sum decomposition $W_+ = \Theta \oplus K^{-1}$. Hence

$$S_\phi(A, \psi) = \int_M \left[|\nabla^A \alpha|^2 + |\nabla^A \beta|^2 + \frac{s}{4}(|\alpha|^2 + |\beta|^2) + |(da)^+ - \omega|^2 \right.$$

$$\left. + |\sigma(\psi)|^2 + 2\langle \sigma(\psi), F_{A_0}^+ + \omega \rangle \right] dV.$$

Using the formula

$$\sigma(\psi) = i \begin{pmatrix} |\alpha|^2 - |\beta|^2 & 2\bar{\alpha}\beta & 0 & 0 \\ 2\alpha\bar{\beta} & |\beta|^2 - |\alpha|^2 & 0 & 0 \\ 0 & 0 & 0 & 0 \\ 0 & 0 & 0 & 0 \end{pmatrix}, \tag{3.29}$$

together with the fact that $F_{A_0}^+$ and ω are represented by diagonal matrices, we can easily check that every term in the last expression for $S_\phi(A, \psi)$ is invariant under the transformation (3.28).

We now focus on the second of the Seiberg-Witten equations,

$$(da)^+ - \omega = \sigma(\psi).$$

Since (3.28) leaves $(da)^+$ alone but changes the signs of the off-diagonal terms in $\sigma(\psi)$, we see that either $\alpha = 0$ or $\beta = 0$. Comparison of (3.29) with the Clifford algebra expression for the Kähler form,

$$\omega = e_1 \cdot e_2 + e_3 \cdot e_4 = \begin{pmatrix} -2i & 0 & 0 & 0 \\ 0 & 2i & 0 & 0 \\ 0 & 0 & 0 & 0 \\ 0 & 0 & 0 & 0 \end{pmatrix},$$

shows that

$$\sigma(\psi) \wedge \omega = \frac{1}{2}(|\beta|^2 - |\alpha|^2)\omega \wedge \omega.$$

Since F_A and F_{A_0} lie in the same de Rham cohomology class, and

$$F_A \wedge \omega = F_A^+ \wedge \omega, \qquad F_{A_0} \wedge \omega = F_{A_0}^+ \wedge \omega,$$

we see that

$$0 = \int_M (F_A^+ - F_{A_0}^+) \wedge \omega = \int_M (da)^+ \wedge \omega = \int_M \left(1 + \frac{1}{2}(|\beta|^2 - |\alpha|^2)\right) \omega \wedge \omega.$$

Hence

$$\int_M (|\alpha|^2 - |\beta|^2)\omega \wedge \omega > 0,$$

it is β which must be zero, and

$$(da)^+ = [1 - (1/2)|\alpha|^2]\omega.$$

Equation (3.25) from the previous section shows that

$$\sqrt{2}[\bar{\partial}\alpha - (ia)_{(0,1)} \cdot \alpha] = D_A^+ \begin{pmatrix} \alpha \\ 0 \end{pmatrix} = 0,$$

so that our perturbed equations become

$$\bar{\partial}(\log \alpha) = (ia)_{(0,1)}, \qquad (da)^+ = [1 - (1/2)|\alpha|^2]\omega, \qquad \delta a = 0,$$

the last being the familiar gauge condition.

Finally, a short calculation shows that

$$\Delta(\log |\alpha|^2) = 2\,\mathrm{Re}(\Delta \log \alpha)$$

$$= -2\,\mathrm{Re}\langle \omega, i\partial\bar{\partial}(\log \alpha)\rangle = 2\,\mathrm{Re}\langle \omega, \partial(a_{(0,1)})\rangle.$$

On the other hand, since ω is self-dual and a is a real one-form,

$$\langle \omega, (da)^+ \rangle = \langle \omega, da \rangle = \langle \omega, \partial(a_{(0,1)}) + \bar{\partial}(a_{(1,0)})\rangle = 2\langle \omega, \partial(a_{(0,1)})\rangle.$$

Hence

$$\Delta(\log |\alpha|^2) = \langle \omega, (da)^+ \rangle = [1 - (1/2)|\alpha|^2]|\omega|^2 = 2 - |\alpha|^2.$$

Recall that by our conventions on the sign of the Laplacian, $\Delta f(p) \geq 0$ if p is a local maximum for f. Hence if $|\alpha| > \sqrt{2}$ or $|\alpha| < \sqrt{2}$ on an open set, the maximum principle yields a contradiction. It follows that $|\alpha| \equiv \sqrt{2}$, $(da)^+ \equiv 0$ and $F_A^+ = F_{A_0}^+$.

Since $b_1 = 0$, it follows from Hodge theory that

$$(da)^+ = 0 = \delta a \quad \Rightarrow \quad a = 0, \quad \text{and hence} \quad \bar{\partial}(\log \alpha) = 0.$$

Thus α is holomorphic, and it follows from the maximum modulus principle that α is constant. Hence $A = A_0$, while α is uniquely determined up to multiplication by a complex number of length one. After dividing by the constant gauge transformations, we obtain a moduli space which consists of a single point.

To finish the argument, we need to check that the linearized equations

$$D_A \psi' - ia' \cdot \psi = 0, \qquad (da')^+ = \sigma(\psi', \psi) + \sigma(\psi, \psi'), \qquad \delta a' = 0$$

are nondegenerate at this solution. To do this, we consider the second derivative of the functional $S_\phi : \mathcal{A} \to \mathbb{R}$ at the solution $(A_0, (\sqrt{2}, 0))$, which minimizes S_ϕ. Solutions to the first two linearized equations are solutions to the variational equation

$$d^2 S_\phi(A, \psi)((a', \psi'), (a', \psi')) = 0,$$

an equation which is invariant under the linearization of (3.28) at the solution $(A_0, (\sqrt{2}, 0))$:

$$a' \mapsto a', \qquad \alpha' \mapsto \alpha', \qquad \beta' \mapsto -\beta' \tag{3.30}$$

At our solution, the second of the linearized equations becomes

$$(da')^+ = 4i \text{ Trace-free Hermitian part of } \begin{pmatrix} \sqrt{2} \\ 0 \end{pmatrix} (\alpha' \quad \beta')$$

$$= \sqrt{2}i \begin{pmatrix} \alpha' + \bar{\alpha}' & 2\beta' \\ 2\bar{\beta}' & -\alpha' - \bar{\alpha}' \end{pmatrix}.$$

Since the transformation (3.30) takes solutions to solutions, $\beta' = 0$. Similarly, since the transformation

$$a' \mapsto a', \qquad \alpha' \mapsto -\alpha', \qquad \beta' \mapsto \beta'$$

takes solutions to solutions, α' is purely imaginary. The last two linearized equations now imply that $(da')^+ = 0$, $\delta a' = 0$, so a' is harmonic, and since $b_1 = 0$, $a' = 0$. The first of the linearized equations finally implies that α' is an imaginary constant, hence generated by an infinitesimal gauge transformation, and the theorem is proven.

Remark: The theorem of Witten was extended by Taubes to symplectic manifolds with K taken to be the canonical bundle for a compatible almost complex structure. A particularly nice treatment of this extension is presented on pages 59-61 of [13].

Corollary. *A Kähler manifold with $b_1 = 0$ and $b^+ \geq 2$ cannot be decomposed into a smooth connected sum of $P^2\mathbb{C}$'s and $\overline{P^2\mathbb{C}}$'s.*

Proof: As we have seen before, a connected sum of $P^2\mathbb{C}$'s and $\overline{P^2\mathbb{C}}$'s can be given a metric of positive scalar curvature by the construction given in [20], and must therefore have vanishing Seiberg-Witten invariants. However the preceding theorem shows that a Kähler manifold with $b_1 = 0$ and $b^+ \geq 2$ has nonvanishing Seiberg-Witten invariants.

This corollary allows us to construct many pairs of smooth manifolds which are homeomorphic but not diffeomorphic. For example, by the discussion in §3.7, a complex surface in $P^3\mathbb{C}$ defined by a nonsingular homogeneous polynomial of fifth degree is a simply connected Kähler manifold which has odd intersection form, and

$$b_+ = 9, \qquad b_- = 44.$$

Hence by the classification of odd quadratic forms, its intersection form is

$$Q = 9(1) \oplus 44(-1),$$

and by Freedman's theory it is also homeomorphic to $9(P^2\mathbb{C})\sharp44(\overline{P^2\mathbb{C}})$. However, it has nonvanishing Seiberg-Witten invariants, so it cannot be diffeomorphic to $9(P^2\mathbb{C})\sharp44(\overline{P^2\mathbb{C}})$. More generally, each of the hypersurfaces M_n of odd degree n, described in §3.7, is homeomorphic but not diffeomorphic to a connected sum of $P^2\mathbb{C}$'s and $\overline{P^2\mathbb{C}}$'s. Thus we get many pairs of compact simply connected four-manifolds which are homeomorphic but not diffeomorphic, corroborating Theorem B from §1.1.

In the earlier technology, the above corollary would have been obtained via Donaldson's polynomial invariants constructed via nonabelian Yang-Mills theory[12].

Actually, the first examples provided by Donaldson of compact simply connected four-manifolds that are homeomorphic but not diffeomorphic satisfy the condition $b_+ = 1$. The Seiberg-Witten theory allows a very simple treatment of these examples, based upon the following proposition.

Proposition. *Suppose that a simply connected Kähler manifold with $b^+ = 1$ has a canonical bundle K which satisfies the conditions*

$$\langle c_1(K)^2, [M] \rangle \geq 0, \qquad \langle c_1(K) \cup [\omega], [M] \rangle > 0. \qquad (3.31)$$

Then $SW(K^{-1/2})$ is well-defined and $SW(K^{-1/2}) = \pm 1$.

Proof: Note that according to the discussion in §3.7, the first of these hypotheses on the canonical bundle, together with the fact that the canonical bundle is not trivial, imply that the Seiberg-Witten invariant $SW(K^{-1/2})$ is well-defined (when the perturbation ϕ is chosen to be small) even though $b_+ = 1$. The proof of the previous theorem shows that the moduli space of solutions to the Seiberg-Witten equations (3.27) when $\phi = F_{A_0}^+ + \omega$ consists of a single nondegenerate point. To see that the same conclusion holds when ϕ is small, we need to check that there are no singular solutions to the Seiberg-Witten equations with $\psi = 0$ when $\phi = t(F_{A_0}^+ + \omega)$, for any $t \in (0,1)$. But such a solution would imply

$$F_A^+ = t(F_{A_0}^+ + \omega) \ \Rightarrow \ (1-t)\int_M (F_A^+) \wedge \omega > 0 \ \Rightarrow \ \int_M (F_A) \wedge \omega > 0.$$

This contradicts the second hypothesis since $c_1(K) = -c_1(K^{-1}) = -2[F_A]$.

To describe Donaldson's examples, we need to utilize some results from the theory of algebraic surfaces which can be found in [19]. Let f and g be two generic cubic homogeneous polynomials in three variables, which define a map

$$[f,g] : P^2 \mathbb{C} \to P^1 \mathbb{C}, \qquad [f,g]([z_0, z_1, z_2]) = [f(z_0, z_1, z_2), g(z_0, z_1, z_2)],$$

except at nine points of $P^2\mathbb{C}$ at which f and g have common zeros. (In the language of algebraic geometry, $[f,g]$ is a *rational map*; see [19], page 491 for the definition.) The preimage of a generic point in $P^1\mathbb{C}$ via $[f,g]$ is a Riemann surface of genus one, defined by a cubic polynomial equation $f + \lambda g = 0$.

To handle the nine singularities of $[f,g]$, we use the notion of blow-up of an algebraic surface ([19], page 182). Blowing M up at the nine common zeros of f and g yields a new algebraic surface Y which is diffeomorphic to $P^2\mathbb{C} \sharp (9\overline{P^2\mathbb{C}})$ and a genuine holomorphic map $Y \to P^1\mathbb{C}$ in which the preimage of a generic point is still a torus. Y is an example of an elliptic surface as described in [19], pages 564-572. Acording to a general formula for the canonical bundle of an elliptic surface ([19], page 572),

$$c_1(K) = \mu(-[F]),$$

where $[F]$ is the homology class of the fiber of $Y \to P^1\mathbb{C}$ and μ is the Poincaré duality isomorphism described at the beginning of §3.5. Since two distinct fibers have empty intersection and the integral of the Kähler form over a fiber must be positive,

$$\langle c_1(K)^2, [M] \rangle = 0, \qquad \langle c_1(K) \cup [\omega], [M] \rangle < 0.$$

The last inequality is reassuring, since $P^2\mathbb{C}\sharp(9\overline{P^2\mathbb{C}})$ has a metric of positive scalar curvature, as we have seen, so $SW(K^{-1/2})$ must vanish.

However, we can modify the sign of the canonical bundle by performing appropriate logarithmic transformations on the fibers of F, thereby obtaining a manifold which satisfies the hypotheses of the proposition. The logarithmic transformation, described in [19], page 566, excises a tubular neighborhood of a smooth fiber and glues it back in with a new diffeomorphism along the boundary, thereby obtaining a new elliptic surface. We can perform two logarithmic transformations on Y obtaining a new algebraic surface Z, with holomorphic map $Z \to P^1\mathbb{C}$, such that two of the fibers (which we denote by F_2 and F_3) have multiplicities two and three, and hence

$$[F] = 2[F_2] = 3[F_3].$$

The new surface Z is called a *Dolgachev surface*. We can still apply the formula for $c_1(K)$, but this time we obtain

$$c_1(K) = \mu(-[F] + [F_2] + 2[F_3]) = \mu((1/6)[F]).$$

Since the integral of the Kähler form over F is positive, (3.31) holds.

Dolgachev showed that Z is simply connected, and it is possible to verify that Y and Z have the same intersection forms (see [11]). Thus by Freedman's theory, Y and Z are homeomorphic. However, the proposition implies that $SW(K^{-1/2}) = \pm 1$, and hence Y and Z are not diffeomorphic.

Using more general logarithmic transformations, one can construct elliptic surfaces homeomorphic to $P^2\mathbb{C}\sharp9(\overline{P^2\mathbb{C}})$ in which $c_1(K)$ satisfies (3.31) and is divisible by an arbitrarily large integer. (See Proposition 3.7 on page 316 of [18].) Given any smooth four-manifold, inequalities (3.5) and (3.13) give a bound on how divisible $c_1(L)$ can be if $SW(L)$ is defined and nonzero. Any finite collection of smooth manifolds is also subject to such a bound. This fact yields a simple proof of the following striking theorem:

Theorem [18], [34]. *The compact topological manifold* $P^2\mathbb{C}\sharp9\overline{P^2\mathbb{C}}$ *has infinitely many distinct smooth structures.*

Bibliography

[1] S. Akbulut, *Lectures on Seiberg-Witten invariants*, Proc. 1995 Gökova Geometry-Topology Conference, Scientific and Technical Research Council of Turkey, 1996, 95-118.

[2] N. Aronszahn, *A unique continuation theorem for solutions of elliptic partial differential equations or inequalities of the second order*, J. Math. Pure Appl. **36** (1957), 235-249.

[3] M. F. Atiyah, *K-theory*, Benjamin, New York, 1967.

[4] M. F. Atiyah, R. Bott and A. Shapiro *Clifford modules*, Topology **3**, Suppl. I (1964), 3-38.

[5] W. Barth, C. Peters and A. Van de Ven, *Compact complex surfaces*, Springer, New York, 1984.

[6] N. Berline, E. Getzler and M. Vergne, *Heat kernels and Dirac operators*, Springer, New York, 1991.

[7] B. Booss-Bavnek and K. P. Wojciechowski, *Elliptic boundary problems for Dirac operators*, Birkhäuser, Boston, 1993.

[8] R. Bott and L. Tu, *Differential forms and algebraic topology*, Springer, New York, 1982.

[9] P. Deligne, P. Etinghof, D. Freed, L. Jeffrey, D. Kazhdan, J. Morgan, D. Morrison and E. Witten, *Quantum field theory and strings: a course for mathematicians*, 2 volumes, American Mathematical Society, Providence, 1999.

[10] S. Donaldson, *An application of gauge theory to four-dimensional topology*, J. Differential Geometry **18** (1983), 279-315.

[11] S. Donaldson, *Irrationality and the h-cobordism conjecture*, J. Differential Geometry **26** (1987), 141-168.

[12] S. Donaldson, *Polynomial invariants for smooth 4-manifolds*, Topology **29** (1990), 257-315.

[13] S. Donaldson, *The Seiberg-Witten equations and 4-manifold topology*, Bull. Amer. Math. Soc. **33** (1996), 45-70.

[14] S. Donaldson and P. B. Kronheimer, *The geometry of four-manifolds*, Oxford Univ. Press, Oxford, 1984.

[15] N. Elkies, *A characterization of the Z^n lattice*, Math. Research Letters **2** (1995), 321-326.

[16] D. Freed and K. Uhlenbeck, *Instantons and four-manifolds*, Springer, New York, 1990.

[17] M. Freedman and F. Quinn, *Topology of 4-manifolds*, Princeton Univ. Press, Princeton, New Jersey, 1990.

[18] R. Friedman and J. W. Morgan, *On the diffeomorphism type of certain algebraic surfaces I, II*, J. Differential Geometry **27** (1988), 297-369.

[19] P. Griffiths and J. Harris, *Principles of algebraic geometry*, Wiley, New York, 1978.

[20] M. Gromov and H. B. Lawson, *The classification of simply connected manifolds of positive scalar curvature*, Annals of Math. **111** (1980), 423-434.

[21] V. Guillemin and A. Pollack, *Differential topology*, Prentice-Hall, New York, 1974.

[22] J. Kollar, *The structure of algebraic three-folds: an introduction to Mori's program*, Bull. Amer. Math. Soc. **17** (1987), 211-273.

[23] P. B. Kronheimer and T. S. Mrowka, *The genus of embedded surfaces in the projective plane*, Math. Research Letters **1** (1994), 797-808.

[24] S. Lang, *Differential and Riemannian manifolds*, Springer Verlag, New York, 1995.

[25] H. B. Lawson and M. L. Michelsohn, *Spin geometry*, Princeton univ. Press, Princeton, New Jersey, 1989.

[26] V. Mathai and D. Quillen, *Superconnections, Thom classes, and equivariant differential forms*, Topology **25** (1986), 85-110.

[27] D. McDuff and D. Salamon, *J-holomorphic curves and quantum cohomology*, American Mathematical Society, Providence, Rhode Island, 1994.

[28] J. Milnor, *Morse theory*, Princeton Univ. Press, Princeton, New Jersey, 1963.

[29] J. Milnor and D. Husemoller, *Symmetric bilinear forms*, Springer, New York, 1973.

[30] J. Milnor and J. Stasheff, *Characteristic classes*, Princeton Univ. Press, Princeton, New Jersey, 1974.

[31] C. W. Misner, K. Thorne and J. A. Wheeler, *Gravitation*, W. H. Freeman, New York, 1973.

[32] J. Morgan, *The Seiberg-Witten equations and applications to the topology of smooth four-manifolds*, Princeton Univ. Press, Princeton, New Jersey, 1996.

[33] J. Morgan, Z. Szabo and C. Taubes, *A product formula for the Seiberg-Witten invariants and the generalized Thom conjecture*, J. Differential Geometry **44** (1996), 706-788.

[34] C. Okonek and A. Van de Ven, *Stable vector bundles and differentiable structures on certain elliptic surfaces*, Inventiones Math. **86** (1986), 357-370.

[35] M. Peskin and D. Schroeder, *An introduction to quantum field theory*, Perseus Books, Reading, Mass. 1995.

[36] S. Smale, *An infinite-dimensional version of Sard's theorem*, Amer. J. Math. **87** (1966), 861-866.

[37] N. Steenrod, *The topology of fiber bundles*, Princeton Univ. Press, Princeton, New Jersey, 1951.

[38] C. Taubes, *The Seiberg-Witten and Gromov invariants*, Math. Research Letters **2** (1995), 221-238.

[39] F. W. Warner, *Foundations of differentiable manifolds and Lie groups*, Scott-Foresman, Glenview, Illinois, 1971.

[40] G. W. Whitehead, *Elements of homotopy theory*, Springer Verlag, New York, 1978.

[41] E. Witten, *Monopoles and four-manifolds*, Math. Research Letters **1** (1994), 769-796.

Index

Printing: Weihert-Druck GmbH, Darmstadt
Binding: Buchbinderei Schäffer, Grünstadt

Recent Reprints and New Editions